MCAT® Physics

2025–2026 Edition: An Illustrated Guide

Copyright © 2024
On behalf of UWorld, LLC
Dallas, TX
USA

All rights reserved.
Printed in English, in the United States of America.

Reproduction or translation of any part of this work beyond that permitted by Sections 107 and 108 of the United States Copyright Act without the permission of the copyright owner is unlawful.

The Medical College Admission Test (MCAT®) and the United States Medical Licensing Examination (USMLE®) are registered trademarks of the Association of American Medical Colleges (AAMC®). The AAMC® neither sponsors nor endorses this UWorld product.

Facebook® and Instagram® are registered trademarks of Facebook, Inc. which neither sponsors nor endorses this UWorld product.

X is an unregistered mark used by X Corp, which neither sponsors nor endorses this UWorld product.

Acknowledgments for the 2025–2026 Edition

Ensuring that the course materials in this book are accurate and up to date would not have been possible without the multifaceted contributions from our team of content experts, editors, illustrators, software developers, and other amazing support staff. UWorld's passion for education continues to be the driving force behind all our products, along with our focus on quality and dedication to student success.

About the MCAT Exam

Taking the MCAT is a significant milestone on your path to a rewarding career in medicine. Scan the QR codes below to learn crucial information about this exam as you take your next step before medical school.

Basic MCAT Exam Information

Scores and Percentiles

MCAT Sections

Registration Guide

Preparing for the MCAT with UWorld

The MCAT is a grueling exam spanning seven subjects that is designed to test your aptitude in areas essential for success in medicine. Preparing for the exam can be intimidating—so much so that in post-MCAT questionnaires conducted by the AAMC®, a majority of students report not feeling confident about their MCAT performance.

In response, UWorld set out to create premier learning tools to teach students the entire MCAT syllabus, both efficiently and effectively. Taking what we learned from helping over 90% of medical students prepare for their medical board exams (USMLE®), we launched the UWorld MCAT Qbank in 2017 and the UWorld MCAT UBooks in 2024. The MCAT UBooks are meticulously written and designed to provide you with the knowledge and strategies you need to meet your MCAT goals with confidence and to secure your future in medical school.

Below, we explain how to use the MCAT UBooks and MCAT Qbank together for a streamlined learning experience. By strategically integrating both resources into your study plan, you will improve your understanding of key MCAT content as well as build critical reasoning skills, giving you the best chance at achieving your target score.

MCAT UBooks: Illustrated and Annotated Guides

The MCAT UBooks include not only the printed editions for each MCAT subject but also provide digital access to interactive versions of the same books. There are eight printed MCAT UBooks in all, six comprehensive review books covering the science subjects and two specialized books for the Critical Analysis and Reasoning Skills (CARS) section of the exam:

- Biology
- Biochemistry
- General Chemistry
- Organic Chemistry
- Physics
- Behavioral Sciences
- CARS (Annotated Practice Book)
- CARS Passage Booklet (Annotated)

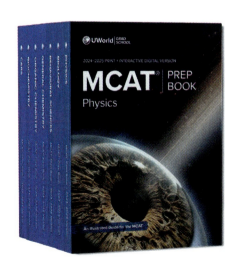

Each UBook is organized into Units, which are divided into Chapters. The Chapters are then split into Lessons, which are further subdivided into Concepts.

MCAT Sciences: Printed UBook Features

The MCAT UBooks bring difficult science concepts to life with thousands of engaging, high-impact visual aids that make topics easier to understand and retain. In addition, the printed UBooks present key terms in blue, indicating clickable illustration hyperlinks in the digital version that will help you learn more about a scientific concept.

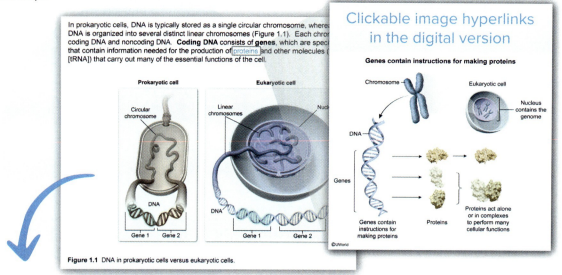

Thousands of educational illustrations in the print book

Clickable image hyperlinks in the digital version

Test Your Basic Science Knowledge with Concept Check Questions

The printed UBooks also include 450 new questions—never before available in the UWorld Qbank—for Biology, General Chemistry, Organic Chemistry, Biochemistry, and Physics. These new questions, called Concept Checks, are interspersed throughout the entire book to enhance your learning experience. Concept Checks allow you to instantly test yourself on MCAT concepts you just learned from the UBook.

Short answers to the Concept Checks are found in the appendix at the end of each printed UBook. In addition, the digital version of the UBook provides an interactive learning experience by giving more detailed, illustrated, step-by-step explanations of each Concept Check. These enhanced explanations will help reinforce your learning and clarify any areas of uncertainty you may have.

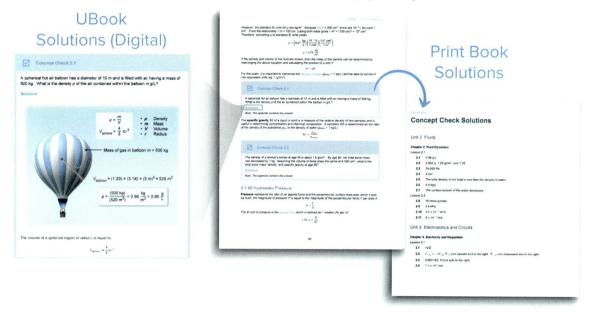

UBook Solutions (Digital)

Print Book Solutions

MCAT CARS Printed UBook Features

For CARS, the main book, or Annotated Practice Book, teaches you the specialized CARS skills and strategies you need to master and then follows up with multiple sets of MCAT-level practice questions.

Additionally, the CARS Passage Booklet includes annotated versions of the passages in the CARS Main Book. From these annotations, you will learn how to break down a CARS passage in a step-by-step manner to find the right answer to each CARS question.

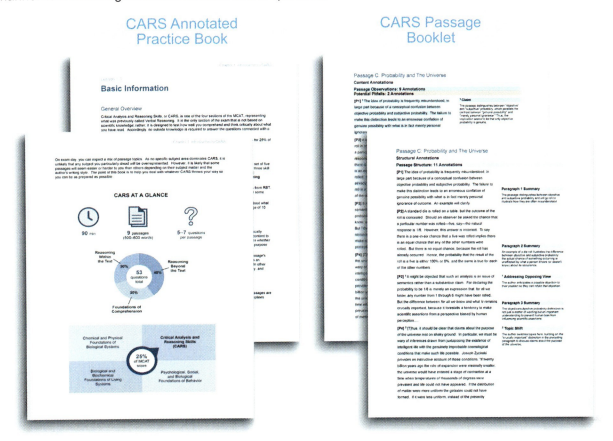

CARS Annotated Practice Book

CARS Passage Booklet

MCAT-Level Exam Practice with the UWorld Qbank

UWorld's MCAT UBooks and Qbank were designed to be used together for a comprehensive review experience. The UWorld Qbank provides an active learning approach to MCAT prep, with thousands of MCAT-level questions that align with each UBook.

The printed UBooks include a prompt at the end of each unit that explains how to access unit practice tests in the MCAT Qbank. In addition, the MCAT UBooks' digital platform enables you to easily create your own unit tests based on each MCAT subject.

To purchase MCAT Qbank access or to begin a free seven-day trial, visit gradschool.uworld.com/mcat.

Boost Your Score with the #1 MCAT Qbank

Scan for free trial

Why use the UWorld Qbank?

- Thousands of high-yield MCAT-level questions
- In-depth, visually engaging answer explanations
- Confidence-building user interface identical to the exam
- Data-driven performance and improvement tracking
- Fully featured mobile app for on-the-go review

Special Features Integrating Digital UBooks and the UWorld Qbank

The digital MCAT UBooks and the MCAT Qbank come with several integrated features that transform ordinary reading into an interactive study session. These time-saving tools enable you to personalize your MCAT test prep, get the most out of our detailed explanations, save valuable time, and know when you are ready for exam day.

My Notebook

My Notebook, a personalized note-taking tool, allows you to easily copy and organize content from the UBooks and the Qbank. Simplify your study routine by efficiently recording the MCAT content you will encounter in the exam, and streamline your review process by seamlessly retrieving high-yield concepts to boost your study performance—in less time.

Digital Flashcards

Our unique flashcard feature makes it easy for students to copy definitions and images from the MCAT UBooks and Qbank into digital flashcards. Each card makes use of spaced repetition, a research-supported learning methodology that improves information retention and recall. Based on how you rate your understanding of flashcard content, our algorithm will display the card more or less frequently.

Fully Featured Mobile App

Study for your MCAT exams anytime, anywhere, with our industry-leading mobile app that provides complete access to your MCAT prep materials and that syncs seamlessly across all devices. With the UWorld MCAT app, you can catch up on reading, flip through flashcards between classes, or take a practice quiz during lunch to make the most of your downtime and keep MCAT material top of mind.

Book and Qbank Progress Tracking

Track your progress while using the MCAT UBooks and Qbank, and review MCAT content at your own pace. Our learning tools are enhanced by advanced performance analytics that allow users to assess their preparedness over time. Hone in on specific subjects, foundations, and skills to iron out any weaknesses, and even compare your results with those of your peers.

Explore the Periodic Table

You will need to use the periodic table to answer questions on the MCAT for specific sections. Introductory general chemistry concepts constitute 30% of the material tested in the Chemical and Physical Foundations of Biological Systems section of the exam. In addition, General Chemistry constitutes 5% of the Biological and Biochemical Foundations of Living Systems section of the MCAT. Using and understanding the periodic table is a crucial skill needed for success in these sections.

1	2	3	4	5	6	7	8	9	10	11	12	13	14	15	16	17	18
1 H 1.0																	2 He 4.0
3 Li 6.9	4 Be 9.0											5 B 10.8	6 C 12.0	7 N 14.0	8 O 16.0	9 F 19.0	10 Ne 20.2
11 Na 23.0	12 Mg 24.3											13 Al 27.0	14 Si 28.1	15 P 31.0	16 S 32.1	17 Cl 35.5	18 Ar 39.9
19 K 39.1	20 Ca 40.1	21 Sc 45.0	22 Ti 47.9	23 V 50.9	24 Cr 52.0	25 Mn 54.9	26 Fe 55.8	27 Co 58.9	28 Ni 58.7	29 Cu 63.5	30 Zn 65.4	31 Ga 69.7	32 Ge 72.6	33 As 74.9	34 Se 79.0	35 Br 79.9	36 Kr 83.8
37 Rb 85.5	38 Sr 87.6	39 Y 88.9	40 Zr 91.2	41 Nb 92.9	42 Mo 95.9	43 Tc (98)	44 Ru 101.1	45 Rh 102.9	46 Pd 106.4	47 Ag 107.9	48 Cd 112.4	49 In 114.8	50 Sn 118.7	51 Sb 121.8	52 Te 127.6	53 I 126.9	54 Xe 131.3
55 Cs 132.9	56 Ba 137.3	57 La* 138.9	72 Hf 178.5	73 Ta 180.9	74 W 183.9	75 Re 186.2	76 Os 190.2	77 Ir 192.2	78 Pt 195.1	79 Au 197.0	80 Hg 200.6	81 Tl 204.4	82 Pb 207.2	83 Bi 209.0	84 Po (209)	85 At (210)	86 Rn (222)
87 Fr (223)	88 Ra (226)	89 Ac+ (227)	104 Rf (261)	105 Db (262)	106 Sg (266)	107 Bh (264)	108 Hs (277)	109 Mt (268)	110 Ds (281)	111 Rg (280)	112 Cn (285)	113 Uut (284)	114 Fl (289)	115 Uup (288)	116 Lv (293)	117 Uus (294)	118 Uuo (294)

*	58 Ce 140.1	59 Pr 140.9	60 Nd 144.2	61 Pm (145)	62 Sm 150.4	63 Eu 152.0	64 Gd 157.3	65 Tb 158.9	66 Dy 162.5	67 Ho 164.9	68 Er 167.3	69 Tm 168.9	70 Yb 173.0	71 Lu 175.0
+	90 Th 232.0	91 Pa (231)	92 U 238.0	93 Np (237)	94 Pu (244)	95 Am (243)	96 Cm (247)	97 Bk (247)	98 Cf (251)	99 Es (252)	100 Fm (257)	101 Md (258)	102 No (259)	103 Lr (260)

Table of Contents

UNIT 1 MECHANICS AND ENERGY

CHAPTER 1 MOTION, FORCE, AND ENERGY 1
- Lesson 1.1 Foundations 3
- Lesson 1.2 Translational Motion 9
- Lesson 1.3 Forces 27
- Lesson 1.4 Equilibrium 47
- Lesson 1.5 Work and Energy 57

UNIT 2 FLUIDS

CHAPTER 2 FLUID DYNAMICS 81
- Lesson 2.1 Hydrostatics 83
- Lesson 2.2 Fluids in Motion 99

UNIT 3 ELECTROSTATICS AND CIRCUITS

CHAPTER 3 ELECTRICITY AND MAGNETISM 113
- Lesson 3.1 Electric Charge and Force 115
- Lesson 3.2 Flowing Charge 125
- Lesson 3.3 Capacitance 139
- Lesson 3.4 Magnetism 149

UNIT 4 LIGHT AND SOUND

CHAPTER 4 WAVES, SOUND, AND LIGHT 159
- Lesson 4.1 Periodic Motion and Waves 161
- Lesson 4.2 Sound 173
- Lesson 4.3 Light 199
- Lesson 4.4 Optical Instruments 217

UNIT 5 THERMODYNAMICS

CHAPTER 5 THERMODYNAMICS AND GASES 239
- Lesson 5.1 Thermodynamic Systems 241
- Lesson 5.2 Thermodynamic Laws 245
- Lesson 5.3 Heat 255
- Lesson 5.4 Phases of Matter 265
- Lesson 5.5 Kinetic Theory of Gases 275

APPENDIX

CONCEPT CHECK SOLUTIONS 283
INDEX 289

Unit 1 Mechanics and Energy

Chapter 1 Motion, Force, and Energy

1.1 Foundations

    1.1.01   Units and Dimensions
    1.1.02   Scalars and Vectors

1.2 Translational Motion

    1.2.01   Distance, Displacement, Speed, and Velocity
    1.2.02   Motion Graphs
    1.2.03   Acceleration
    1.2.04   Linear Kinematics Equations
    1.2.05   Projectile Motion

1.3 Forces

    1.3.01   Inertia and Newton's First Law
    1.3.02   Newton's Second Law
    1.3.03   Newton's Third Law
    1.3.04   Applications of Newton's Laws
    1.3.05   Friction
    1.3.06   Hooke's Law

1.4 Equilibrium

    1.4.01   Center of Mass
    1.4.02   Torque
    1.4.03   Static Equilibrium
    1.4.04   Dynamic Equilibrium

1.5 Work and Energy

    1.5.01   Concept of Work
    1.5.02   Kinetic Energy and the Work-Energy Theorem
    1.5.03   Potential Energy of Systems
    1.5.04   Conservation of Total Energy
    1.5.05   Power
    1.5.06   Mechanical Advantage and Simple Machines

Lesson 1.1

Foundations

Introduction

In this lesson, two foundational concepts needed throughout a course in physics are reviewed. Physics questions frequently involve equations and algebraically solving these equations for certain variables. Therefore, understanding that the variables in physics equations have dimensions represented by a system of units is a useful tool in solving problems on the exam.

In addition, it is important to recognize that two different types of variables may appear in questions. Scalars are quantities such as mass, volume, density, and temperature. Quantities associated with motion and direction—such as velocity, acceleration, and force—are vectors. Analyzing vector quantities involves the use of basic trigonometry.

1.1.01 Units and Dimensions

Physical quantities are expressed in terms of **dimensions**. For example, an object has mass, temperature, and a size given by its length in each direction. Furthermore, the motion of the object is described in terms of its behavior over an interval of time.

These dimensions of mass, temperature, length, and time are expressed in terms of a set of **units**. In contrast, **pure numbers** such as the ratio of an object's length and width, are *dimensionless* and have no units.

A common system of units is the International System of Units (or **SI system**), with base units shown in Table 1.1.

Table 1.1 Base SI units.

Physical quantity	Unit
Length	meter (m)
Mass	kilogram (kg)
Time	second (s)
Temperature	Kelvin (K)
Electric current	ampere (A)
Amount of substance	mole (mol)

Important quantities such as force, energy, and pressure are expressed in terms of derived SI units, which are combinations of the basic SI units.

In physics, very large or very small values of these units are frequently encountered in calculations. To concisely display large or small values, the prefixes shown in Table 1.2 are added to SI units.

Table 1.2 List of prefixes.

Prefix	Abbreviation	Multiplicative value	
		Standard notation	Scientific notation
tera	T	1,000,000,000,000	10^{12}
giga	G	1,000,000,000	10^{9}
mega	M	1,000,000	10^{6}
kilo	k	1,000	10^{3}
deci	d	0.1	10^{-1}
centi	c	0.01	10^{-2}
milli	m	0.001	10^{-3}
micro	μ	0.000001	10^{-6}
nano	n	0.000000001	10^{-9}
pico	p	0.000000000001	10^{-12}

For example, in the case of the dimension of length, the prefixed units of kilometer (km) and nanometer (nm) represent 10^3 meters and 10^{-9} meters, respectively.

> **Concept Check 1.1**
>
> The light emitted by a laser has a wavelength of 600 nm. What is this wavelength in meters?
>
> **Solution**
>
> *Note: The appendix contains the answer.*

Keeping track of the dimensions of any variable X is a useful technique for solving physics problems. In an algebraic equation, the dimensions $[X]$ of the left side must be equal to the dimensions of the right side:

$$X = Y \Rightarrow [X] = [Y]$$

Dimensional analysis refers to the process by which the units of one variable featured in an equation can be determined algebraically using the units of the other variables within the same equation.

> **✓ Concept Check 1.2**
>
> Newton's law of universal gravitation states that the force *F* between two objects is equal to the product of the gravitational constant *G*, the masses of the objects (m_1, m_2), and the inverse of their separation length squared r^2:
>
> $$F = \frac{Gm_1m_2}{r^2}$$
>
> The dimensions of force are the product of mass and length divided by time squared. What are the dimensions of *G*?
>
> **Solution**
>
> *Note: The appendix contains the answer.*

1.1.02 Scalars and Vectors

Scalars are physical quantities characterized by only a size or a magnitude—for example, the amount of time elapsed during an experiment, or the mass of an object. **Vector quantities** are characterized by a *direction* in addition to a magnitude.

The kinematic variables that describe the motion of an object (displacement, velocity, and acceleration, which are discussed in Lesson 1.2) are vector quantities because an object can move in any arbitrary direction in space. Furthermore, vector fields (eg, the electric field) represent the magnitude and direction of a vector at each point in space.

Vectors can be represented in two equivalent ways, as shown in Figure 1.1.

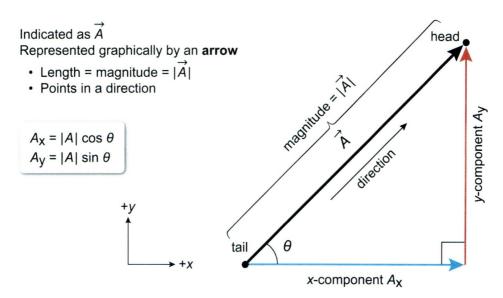

Figure 1.1 Vector characteristics.

One method is to represent the vector with an arrow, which shows the magnitude and direction. The length of the arrow represents the vector's magnitude, and the direction of the arrow is aligned with the direction of the vector.

In the second approach, an *xy*-coordinate system is introduced. The vector (denoted by $\vec{A}$) can then be represented in terms of *x*- and *y*-**components** (A_x, A_y) with respect to this coordinate system:

$$\vec{A} = (A_x, A_y)$$

Furthermore, as shown in Figure 1.1, the vector components form a right triangle inclined at an angle θ above the *x*-axis. The vector magnitude is the hypotenuse of the triangle, and A_x and A_y are the adjacent and opposite legs of the resulting triangle. Hence, trigonometric ratios can be used to express the vector components in terms of the magnitude $|A|$ and the sine and cosine functions:

$$\vec{A} = (|A|\cos\theta, |A|\sin\theta)$$

Moreover, by the Pythagorean theorem, $|A|$ is equal to the square root of the sum of the squares of the *x*- and *y*-components:

$$|A| = \sqrt{A_x^2 + A_y^2}$$

The **addition** of two vectors can be performed either graphically or via components. The sum of two vectors $\vec{A}$ and $\vec{B}$ can be determined using the **tip-to-tail method** shown in Figure 1.2.

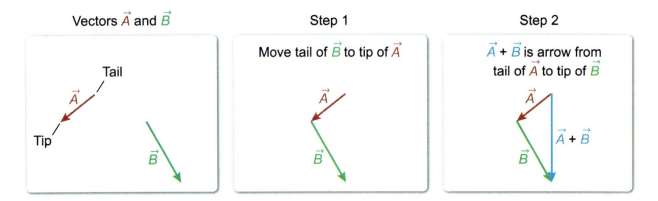

Figure 1.2 Graphical addition of two vectors.

With this method, the tail of the arrow representing $\vec{A}$ is shifted such that it touches the tip (ie, head) of the arrow representing $\vec{B}$. The sum (or resultant vector) $\vec{A} + \vec{B}$ is then represented by the arrow connecting the tail of $\vec{B}$ to the tip of the shifted $\vec{A}$.

In terms of components, vector addition is straightforward. The components of $\vec{A} + \vec{B}$ equal the sum of the components of each vector (see Figure 1.3):

$$\vec{A} + \vec{B} = (A_x + B_x, A_y + B_y)$$

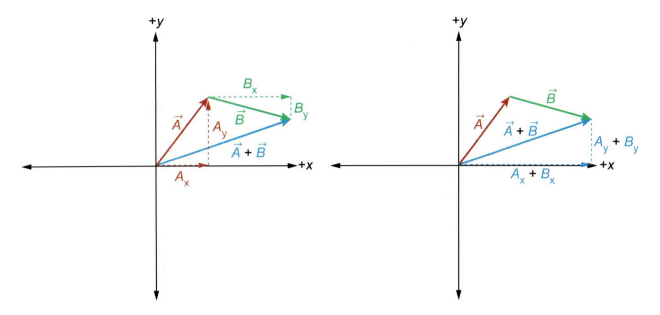

Figure 1.3 Vector addition by components.

☑ Concept Check 1.3

Vector $\vec{A}$ = (10 m, 12 m) and vector $\vec{B}$ = (4 m, 20 m). What is the difference between the two vectors, $\vec{A} - \vec{B}$, and what the magnitude of the difference $|\vec{A} - \vec{B}|$?

Solution
Note: The appendix contains the answer.

Lesson 1.2

Translational Motion

Introduction

This lesson investigates the motion of an object, which can be described by the object's position, velocity, and acceleration. The straight-line motion of an object in one dimension is often expressed graphically in terms of plots of position and velocity as functions of time. This lesson shows how motion variables can be determined using the slope and area under the curve of these graphs.

Furthermore, when an object experiences constant acceleration, the linear kinematics equations can be used to calculate the motion of an object. Finally, this lesson concludes by applying the linear kinematics equations to the case of projectiles launched upward into the air under the influence of gravity.

1.2.01 Distance, Displacement, Speed, and Velocity

The location of an object is given by its **position** $\vec{r}$ (a vector) relative to a particular coordinate system. In terms of *xy*-components:

$$\vec{r} = (x, y)$$

When an object moves from one location to another, its position changes. As shown in Figure 1.4, the object's (ie, the baseball's) **displacement** $\vec{d}$ is the difference between the final and initial positions of the object:

$$\vec{d} = \vec{r}_\text{f} - \vec{r}_\text{i}$$

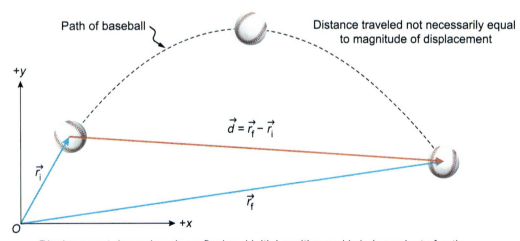

Figure 1.4 Displacement vs. distance.

Displacement depends only on the initial and final position of an object, and therefore does not depend on the path taken between those locations. In contrast, the total **distance** the object travels between the initial and final positions does depend on the path, and the magnitude of displacement is not necessarily equal to the distance traveled:

$$|\vec{d}| \neq \text{Distance}$$

Velocity $\vec{v}$ is a measure of an object's displacement per unit time, hence it is a vector quantity with dimensions of length per time:

$$[v] = \frac{[\text{Length}]}{[\text{Time}]}$$

with the SI units of meters per second (m/s).

Velocity can be measured either as an average over an interval (ie, between two times) or instantaneously. The **average velocity** $\vec{v}_{avg}$ is equal to the change of the object's position over a finite interval Δt:

$$\vec{v}_{avg} = \frac{\Delta \vec{r}}{\Delta t}$$

Figure 1.5 shows a calculation of the baseball's average velocity over an interval.

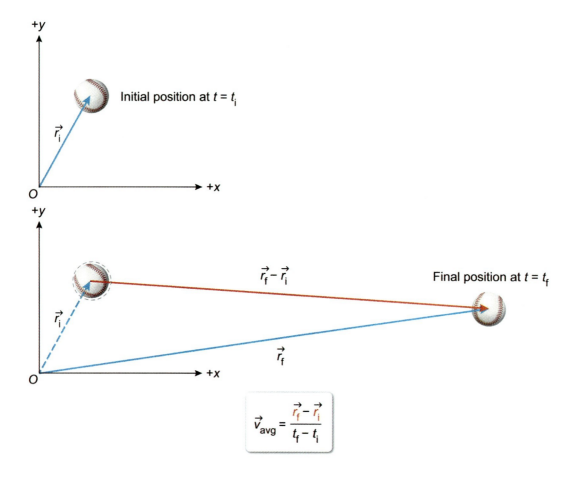

Figure 1.5 Calculating the average velocity of the baseball.

In terms of x- and y-components, the initial and final times and positions can be used:

$$v_{avg,x} = \frac{x_f - x_i}{t_f - t_i} \quad v_{avg,y} = \frac{y_f - y_i}{t_f - t_i}$$

Instantaneous velocity is the velocity of the object at a particular moment in time (ie, t_0) and can be determined by considering the limiting value of the velocity as the interval around t_0 as t goes to zero:

$$\vec{v} = \lim_{\Delta t \to 0} \frac{\Delta \vec{r}}{\Delta t}$$

The instantaneous velocity of the baseball at a point on its path is shown in Figure 1.6.

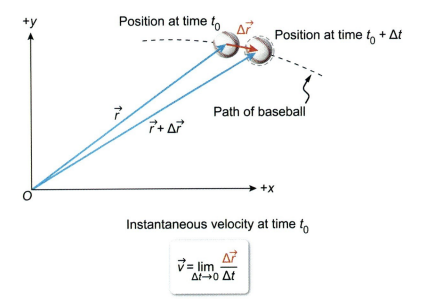

Figure 1.6 Instantaneous velocity.

When the velocity of an object is *constant*, the average velocity is always equal to the instantaneous velocity. Rearranging the definition of velocity implies that displacement is equal to the product of velocity and the interval Δt:

$$\Delta \vec{r} \equiv \vec{d} = \vec{v} \cdot \Delta t$$

Alternatively, when velocity is not constant, the relationship between displacement, velocity, and time is more complicated. The next section discusses how velocity and displacement can be determined via motion graphs of an object's position as a function of time.

Finally, **speed** is the scalar quantity associated with velocity. The **instantaneous speed** s is the magnitude of the instantaneous velocity:

$$s = |\vec{v}|$$

However, the **average speed** s_{avg} of an object is defined as the ratio of the total distance divided by the elapsed time:

$$s_{avg} = \frac{\text{Total distance}}{\text{Elapsed time}}$$

Figure 1.7 shows the average speed of the baseball as it travels a total distance of 20 m between two points in an elapsed time of 2 s.

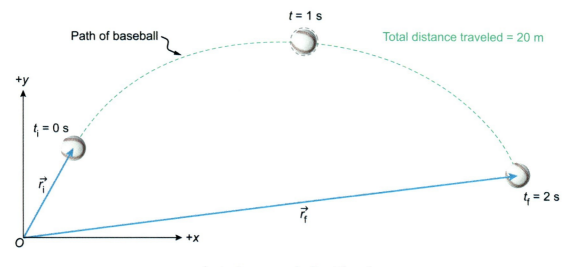

Figure 1.7 Average speed of the baseball.

 Concept Check 1.4

A swimmer swims the full length of a 50 m pool and then returns to the starting point in 50 s. What is (a) the distance traveled by the swimmer, (b) the swimmer's total displacement, (c) the swimmer's average velocity, and (d) the swimmer's average speed?

Solution

Note: The appendix contains the answer.

1.2.02 Motion Graphs

The motion of objects is frequently displayed using graphs of the object's position and velocity as functions of time. In many cases, only motion in only one dimension is considered (ie, back and forth along a straight line).

On a plot of position (vertical axis) versus time (horizontal axis), the slope over a segment with time interval Δt corresponds to the **average velocity**:

$$v_{avg} = \text{slope} = \frac{\text{Rise}}{\text{Run}} = \frac{\Delta x}{\Delta t}$$

In terms of the object's final and initial positions and times:

$$v_{avg} = \frac{x_f - x_i}{t_f - t_i}$$

An example of a position versus time graph is shown in Figure 1.8, where the object shown in the graph has an initial position of 20 m at time $t = 0$. The position of the object then increases to a maximum of 30 m before reaching a final position of 10 m at $t = 8$ s.

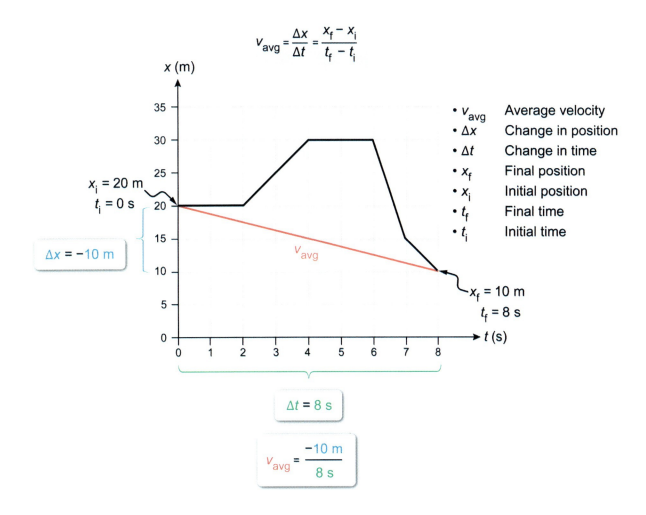

Figure 1.8 Calculating average velocity from a position versus time graph.

Therefore, the average velocity over the 8 s interval equals the slope of the line connecting the initial and final positions on the graph:

$$v_{avg} = \frac{10 \text{ m} - 20 \text{ m}}{8 \text{ s} - 0 \text{ s}} = \frac{-10 \text{ m}}{8 \text{ s}} = -1.25 \frac{\text{m}}{\text{s}}$$

In contrast, the instantaneous slope on a position versus time graph at each point corresponds to the **instantaneous velocity** of the object at a particular moment in time:

$$v = \lim_{\Delta t \to 0} \frac{\Delta x}{\Delta t} = \text{Slope}$$

Examples of instantaneous velocity are shown in Figure 1.9, where the velocity at each time (t_1, t_2, and t_3) is the instantaneous slope of the line tangent to the curve at each point.

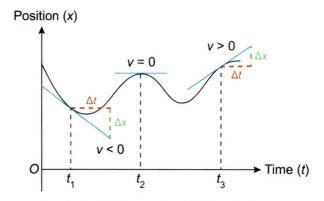

v at given time equals slope of tangent line

Figure 1.9 Velocity from a position vs. time graph.

One can also consider graphs of **velocity versus time**. If the object's velocity as a function of time is known, then its displacement can be determined by rearranging the definition of velocity:

$$\Delta x = v \cdot \Delta t$$

In general, the **area under the curve** of a linear graph in an *xy*-coordinate system is the product of the *x*- and *y*-variables:

$$\text{Area} = yx$$

Hence, displacement is equal to the area under the curve of a velocity versus time graph:

$$\Delta x = d = \text{Area}$$

An example calculation of displacement using a velocity versus time graph is shown in Figure 1.10. The object shown in the graph starts with a velocity of 20 m/s at *t* = 0. The velocity of the object then increases linearly between *t* = 5 s and *t* = 15 s to 40 m/s.

The total displacement of the object from 0 s to 15 s is then determined by calculating the area under the curve, which is the sum of the area of the blue rectangular region A_R and the area of the green triangle A_T:

$$\Delta x = A_R + A_T$$

In terms of the base b and height h of each region:

$$\Delta x = b_R h_R + \frac{1}{2} b_T h_T$$

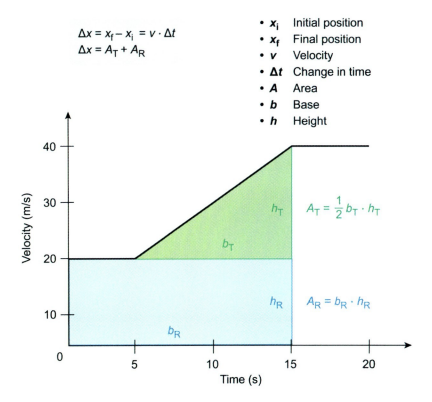

Figure 1.10 Calculating displacement from the area under a curve of velocity versus time.

1.2.03 Acceleration

Just as velocity is the rate of change of position per unit time, **acceleration** is defined as the change in velocity Δv over an interval Δt. Consequently, acceleration has units of length per time squared:

$$[a] = \frac{\frac{\text{Length}}{\text{Time}}}{\text{Time}} = \frac{[\text{Length}]}{[\text{Time}]^2}$$

and can be expressed in SI units of meters per seconds squared (m/s²).

The **average acceleration** a_{avg} measures Δv over a finite Δt and is equal to the ratio of the difference between the final velocity v_f and the initial velocity v_i over the interval:

$$a_{avg} = \frac{\Delta v}{\Delta t} = \frac{v_f - v_i}{t_f - t_i}$$

Instantaneous acceleration a is the value of the acceleration at a particular instant in time in the limit where Δt goes to zero:

$$a = \lim_{\Delta t \to 0} \frac{\Delta v}{\Delta t}$$

Like average and instantaneous velocity, both acceleration concepts can be represented in terms of the slope on a velocity vs. time graph. The object shown in the graph in Figure 1.11 initially has a velocity of 20 m/s at $t = 0$ s. Between 0 s and 8 s, the object's velocity first increases and then decreases to 0 m/s. The a_{avg} over the 8 s interval is the slope of the line between the two points, which equals:

$$a_{avg} = \frac{0\,\frac{m}{s} - 20\,\frac{m}{s}}{8\,s - 0\,s} = \frac{-20\,\frac{m}{s}}{8\,s} = -2.5\,\frac{m}{s^2}$$

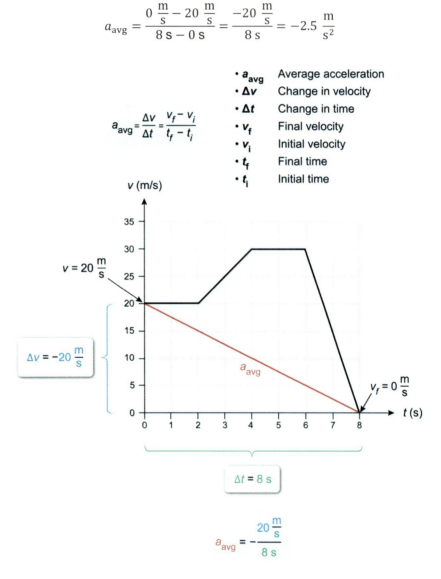

Figure 1.11 Calculating average acceleration from a graph of velocity versus time.

Figure 1.12 depicts the concept of instantaneous acceleration. The acceleration at the marked times on the graph is the slope of the line tangent to the velocity versus time graph at each point. The negative slope at t_1 ($a < 0$) indicates the object is slowing down, and the positive slope at t_3 ($a > 0$) indicates the object is speeding up. When the slope is zero at t_2 ($a = 0$), the object's velocity is momentarily constant.

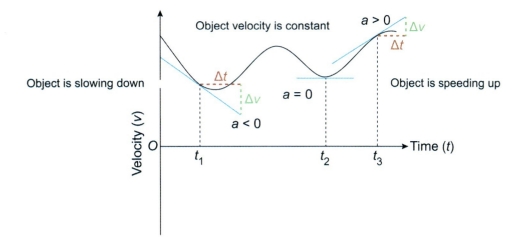

Figure 1.12 Acceleration from a velocity versus time graph.

One can sometimes consider plots of acceleration versus time (which are connected to variable forces, as discussed in Lesson 1.3). The definition of acceleration implies that the **area under the curve** of these graphs is equal to the *change in velocity*:

$$\Delta v = a \cdot \Delta t = \text{Area}$$

Figure 1.13 shows plots of constant, uniform, and non-uniform acceleration versus time, along with the associated areas under each curve.

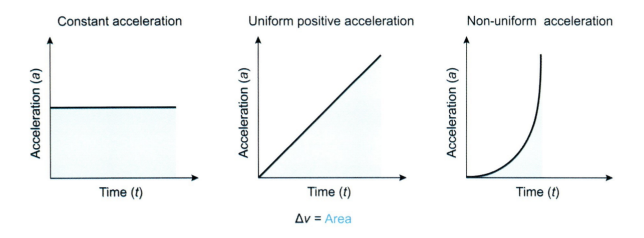

Figure 1.13 Plots of acceleration versus time. The area under the curve is the change in velocity.

Concept Check 1.5

An elevator moves downward from the fourth floor to the first floor of a building. As the elevator approaches the first floor, its motion begins to slow. What are the directions of the elevator's velocity and acceleration?

Solution

Note: The appendix contains the answer.

1.2.04 Linear Kinematics Equations

When an object moves with a constant acceleration a, its position and velocity as functions of time can be determined using the **linear kinematics equations**. The linear kinematics equations are shown in Table 1.3.

Table 1.3 Linear kinematics equations of motion under constant acceleration.

	Equation	Relation
(i)	$v_f = v_i + at$	Velocity with time
(ii)	$v_f^2 = v_i^2 + 2a \cdot \Delta x$	Velocity with displacement
(iii)	$\Delta x = v_i t + \frac{1}{2} at^2$	Displacement with time
(iv)	$\Delta x = v_{avg} \cdot t = \left(\frac{v_i + v_f}{2}\right) \cdot t$	Average velocity

v_i = Initial velocity
v_f = Final velocity
a = Acceleration
t = Time
Δx = Displacement (change in position)
v_{avg} = Average velocity

In the first equation, labeled (i) in Table 1.3, the definition of acceleration is rearranged such that the final velocity is equal to the sum of the initial velocity and the product of a and the elapsed time t. This equation is useful to determine the final velocity of an accelerating object when the object's displacement is unknown or not relevant, as shown in Figure 1.14.

Chapter 1: Motion, Force, and Energy

In this figure, a car with an initial velocity v_i = 5.0 m/s accelerates down an incline with an acceleration a = 2.0 m/s². After 4 s, the equation (i) implies the final velocity v_f equals:

$$v_f = \left(5.0 \frac{m}{s}\right) + \left(2.0 \frac{m}{s^2}\right)(4\ s)$$

$$v_f = 5.0 \frac{m}{s} + 8.0 \frac{m}{s} = 13 \frac{m}{s}$$

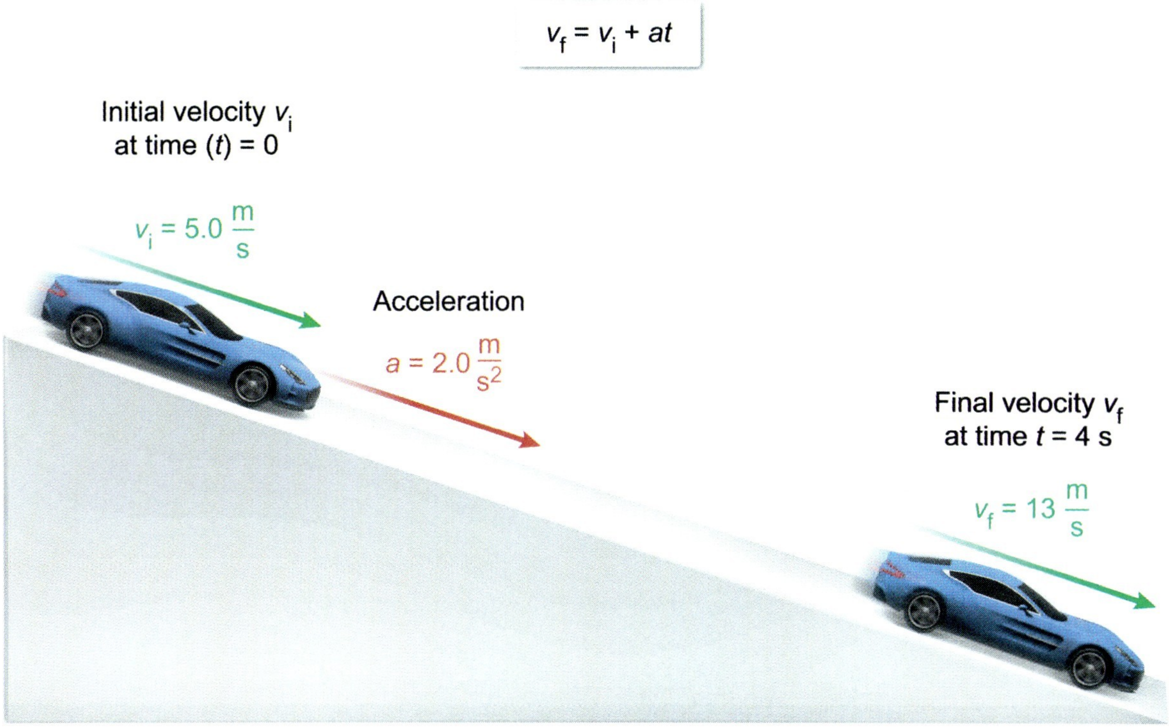

$$v_f = v_i + at$$

Initial velocity v_i
at time (t) = 0

$v_i = 5.0 \frac{m}{s}$

Acceleration

$a = 2.0 \frac{m}{s^2}$

Final velocity v_f
at time t = 4 s

$v_f = 13 \frac{m}{s}$

Figure 1.14 Linear motion of a car with acceleration down an incline.

The second kinematic equation (ii) is useful when the time is unknown or not relevant to the solution. For example, Figure 1.15 shows how the stopping distance of a car can be determined from its initial velocity and (negative) acceleration. The car shown has an initial velocity of 20 m/s and accelerates with a constant value of −4 m/s² as it comes to rest.

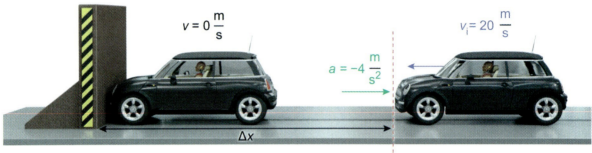

$$v^2 = v_i^2 + 2a \cdot \Delta x$$

$$\Delta x = \frac{v^2 - v_i^2}{2a}$$

- v Final velocity
- v_i Initial velocity
- a Acceleration
- Δx Displacement

$$\Delta x = \frac{\left(0\,\frac{m}{s}\right)^2 - \left(20\,\frac{m}{s}\right)^2}{2\left(-4\,\frac{m}{s^2}\right)}$$

Figure 1.15 Calculating the stopping distance of a car.

The equation implies that as the car brakes, it will take 50 m to completely stop:

$$\Delta x = \frac{\left(0\,\frac{m}{s}\right)^2 - \left(20\,\frac{m}{s}\right)^2}{2\left(-4\,\frac{m}{s^2}\right)} = \frac{-400\,\frac{m^2}{s^2}}{-8\,\frac{m}{s^2}} = 50\text{ m}$$

The third equation (iii) is used to determine displacement given acceleration, initial velocity, and time (ie, the final velocity v_f is unknown or not relevant). In Figure 1.16, the initial position of the rear of the train car is $x_i = -5$ m, and the train moves to the right with an initial velocity $v_i = 2.5$ m/s.

If the train accelerates to the right at 1.0 m/s² for 5 s, the linear kinematic equation implies that the displacement is 25 m:

$$\Delta x = \left(2.5\,\frac{m}{s}\right)(5\text{ s}) + \frac{1}{2}\left(1.0\,\frac{m}{s^2}\right)(5\text{ s})^2$$

$$\Delta x = 12.5\text{ m} + \frac{1}{2}(1.0\text{ m})(25)$$

$$\Delta x = 25\text{ m}$$

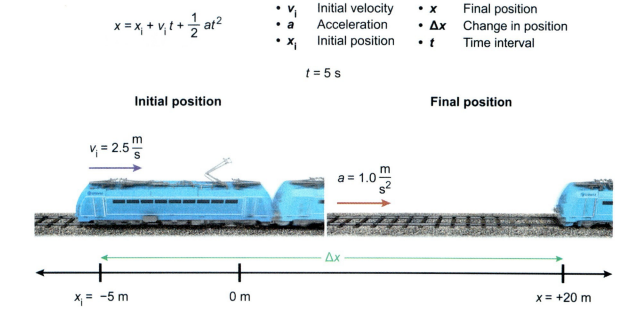

Figure 1.16 Using linear kinematics equations to determine the distance traveled by an accelerating train.

Hence, the final position of the rear of the car is as follows:

$$x_f - x_i = 25 \text{ m}$$
$$x_f = 25 \text{ m} + (-5 \text{ m}) = 20 \text{ m}$$

Finally, the fourth equation (iv) is useful for determining displacement using average velocity when the initial and final velocities of the object are given over a known interval, or when the value of the (constant) acceleration is not needed. For example, if a hockey puck is accelerated from rest to 40 m/s in 0.5 s, the average velocity equation implies that the puck's displacement is equal to 10 m (Figure 1.17):

$$\Delta x = \left(\frac{v_i + v_f}{2}\right) \cdot t$$

$$\Delta x = \left(\frac{0 \, \frac{\text{m}}{\text{s}} + 40 \, \frac{\text{m}}{\text{s}}}{2}\right) \cdot (0.5 \text{ s}) = 10 \text{ m}$$

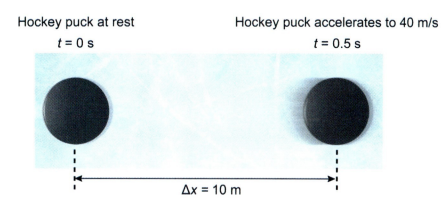

Figure 1.17 Displacement of a uniformly accelerated hockey puck using the average velocity equation.

> ## ✅ Concept Check 1.6
>
> A student drops a pebble from rest down into a deep well. Assuming the pebble's acceleration is 10 m/s² as it falls downward, what one piece of information is needed to determine the depth of the well?
>
> **Solution**
>
> *Note: The appendix contains the answer.*

1.2.05 Projectile Motion

Projectile motion is the movement of objects in two dimensions (i.e., in both the horizontal and vertical directions) under only the effect of gravity. The motion of an object in two dimensions is characterized by its position, velocity, and acceleration vectors as functions of time. Each of these vectors can be resolved into its components (Concept 1.1.02) by introducing an *xy*-coordinate system, as shown in Table 1.4.

Table 1.4 List of components of position, velocity, and acceleration in an *xy*-coordinate system.

	Position	Velocity	Acceleration
x-component	x	v_x	a_x
y-component	y	v_y	a_y

As discussed in Lesson 1.3, the gravitational force experienced by an object (ie, its weight) is associated with a constant vertically downward (−*y* direction) acceleration *g* equal to 9.8 m/s² (Figure 1.18 shows this concept using a baseball):

$$a_y = -g = -9.8 \frac{\text{m}}{\text{s}^2}$$

To simplify calculations, it is often useful to approximate *g* by −10 m/s².

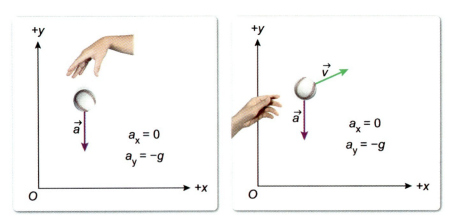

Baseball's acceleration due to gravity is the same regardless of whether it dropped vertically or thrown at an angle

Figure 1.18 Acceleration due to gravity is always vertically downward.

Chapter 1: Motion, Force, and Energy

When assessing projectile motion in two dimensions, the motion of the projectile can be broken into two one-dimensional motions that are independent of each other (Figure 1.19). In the horizontal (+x−direction) the acceleration of a projectile is zero:

$$a_x = 0$$

Hence, the linear kinematics equations shown in Table 1.3 can be applied to analyze the motion of the projectile. In the horizontal direction, zero acceleration means that the projectile's horizontal velocity is always constant:

$$v_x = \text{Constant}$$

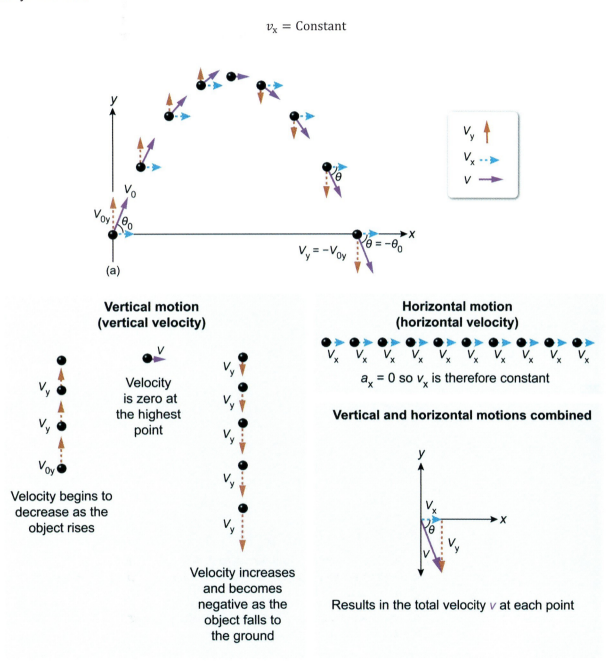

Figure 1.19 Vertical and horizontal velocity in projectile motion

If the object is launched above the horizontal at an angle θ_0 and with initial speed v_i, then trigonometry implies that the horizontal velocity is:

$$v_x = v_i \cos\theta_0 = \text{Constant}$$

However, the object's velocity in the vertical direction changes with time. At first, the initial vertical velocity is:

$$v_{y,i} = v_i \sin\theta_0$$

At later times, the linear kinematics equations yield the following vertical velocity:

$$v_y = v_i \sin\theta_0 - gt$$

Consequently, v_y *decreases with time* to zero and then becomes negative as the object returns to the ground (see Figure 1.19).

The motion of the projectile (ie, a basketball) is shown in Figure 1.20. The highest point of the trajectory shown in the graph is the location where v_y = 0. The maximum height h can be determined using the linear kinematics equation for the initial and final v_y and the vertical displacement Δy:

$$v_{y,f}^2 = v_{y,i}^2 - 2g\Delta y$$

Inserting the values of $v_{y,i}$ and $\Delta y = h$ yields:

$$0 = v_i^2 \sin^2\theta_0 - 2gh$$

Solving for h yields:

$$h = \frac{v_i^2 \sin^2\theta_0}{2g}$$

- v_i Initial velocity
- v_x x-component of velocity
- v_y y-component of velocity
- h Maximum height of projectile
- g Gravitational acceleration

v_x is constant

v_y decreases to zero, then becomes negative

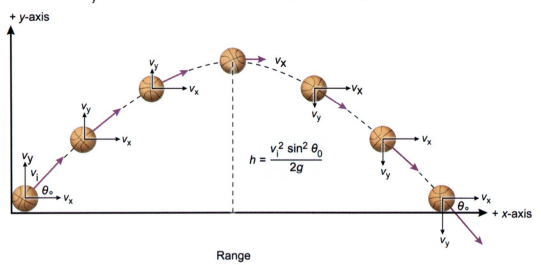

Figure 1.20 Projectile motion of an object under the influence of gravity.

The **range** of the projectile (ie, the horizontal distance traveled) can also be determined using the linear kinematics equations. In summary, the linear kinematics equations can be applied to projectile motion as shown in Table 1.5.

Table 1.5 Linear kinematics equations applied to projectile motion.

Horizontal direction	Vertical direction
v_x = Constant	$v_{y,f} = v_{y,i} - gt$
$\Delta x = v_x t$	$v_{y,f}^2 = v_{y,i}^2 - 2g\Delta y$
	$\Delta y = v_{y,i} t - \frac{1}{2} g t^2$

Time t is the same in both sets of equations

Gravitational acceleration $g = 9.8 \frac{m}{s^2}$

$v_x = v_i \cos \theta_0$

$v_{y,i} = v_i \sin \theta_0$

> ☑ **Concept Check 1.7**
>
> Determine a formula for a projectile's range in terms of v_i, θ_0, and g.
>
> **Solution**
>
> Note: The appendix contains the answer.

Chapter 1: Motion, Force, and Energy

Lesson 1.3

Forces

Introduction

Lesson 1.2 addressed kinematics, the area of physics that describes the motion of objects, such as how fast, how far, or how long it takes an object to move. The area of physics concerned with forces and how they cause motion is called dynamics, which is the subject of this lesson.

1.3.01 Inertia and Newton's First Law

Inertia (meaning "idle" or "sluggish") is the tendency of an object to resist changing its state of motion, and it is a function of the object's **mass**. An object with a large mass has a greater tendency to resist changing its motion than an object with a small mass. For example, dishes at rest on a table remain at rest when the tablecloth is quickly pulled out from underneath, as shown in Figure 1.21.

Tablecloth under dishes   Tablecloth removed quickly   Dishes remain on table

Figure 1.21 Due to inertia, the dishes remain on the table after pulling the tablecloth out.

Alternatively, a space probe moving through deep space continues to do so without the aid of a force. According to the concept of inertia, objects do not require a continuous force to act on them to maintain their present state of motion.

Isaac Newton showed that **forces** are responsible for changing the motion of objects. Force is a vector quantity that has both a **magnitude** and a **direction**, and can be defined simply as a push or a pull. Furthermore, forces can be categorized as **contact** or **field** forces. Contact forces occur when objects come into physical contact (eg, push, pull, tension, friction, normal, spring). Field forces occur when one object exerts an influence on another object without physical contact (eg, gravitational, electromagnetic). Regardless of the nature of the force, the SI unit for force is the newton (N).

In situations where multiple forces act on a given object, the **net force** F_{net} on the object is the vector sum of all the forces acting on it.

$$\vec{F}_{net} = \sum \vec{F}$$

A continuous force is not necessary to keep an object in motion. Objects maintain motion because they have mass (ie, inertia). The hockey puck sliding across the ice seen in Figure 1.22 resists changing its velocity because of its inertia (ie, mass).

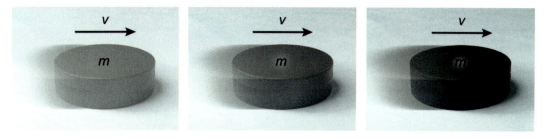

When $F_{net} = 0$, v is constant.
Objects with mass resist changing their velocity

Figure 1.22 The absence of a net force allows a hockey puck to slide across the ice with constant velocity.

This concept of inertia is the basis for **Newton's first law of motion**, which states that an object at rest remains at rest and an object in motion remains in motion with *constant velocity* if the net force acting on the object is equal to zero:

$$\vec{F}_{net} = \sum \vec{F} = 0$$

For example, in Figure 1.23, the crate at rest on a table remains at rest if the downward gravitational force F_g and the upward normal force F_N on the crate are equal (ie, $F_{net} = 0$).

- F_g Gravitational force (weight)
- F_N Normal force

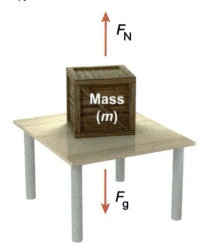

Figure 1.23 The normal and gravitational forces acting on a crate at rest on a table.

For the simple case of a crate at rest on a table, there may only be two forces acting along one dimension. However, when multiple force vectors act in two dimensions, the net force on an object can be found by introducing an *xy*-coordinate system and evaluating the sum of individual forces in the *x*- and *y*-directions:

$$\sum_{i=1}^{n} F_x^{(i)} = F_x^{(1)} + F_x^{(2)} + \cdots + F_x^{(n)} \qquad \sum_{i=1}^{n} F_y^{(i)} = F_y^{(1)} + F_y^{(2)} + \cdots + F_y^{(n)}$$

For example, the airplane shown in Figure 1.24 moves in a straight line with a constant speed. The drag (F_D) and thrust (F_T) forces oppose each other (ie, have opposite signs) in the x-direction, and the lift (F_L) and weight (F_g) oppose each other in the y-direction.

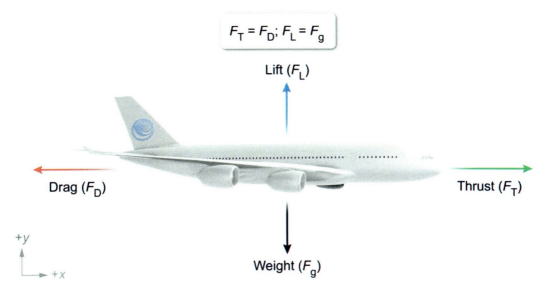

Figure 1.24 An airplane moving with a constant velocity.

Because the airplane is moving with constant velocity (ie, acceleration equals zero), Newton's first law of motion implies that the net vertical force and net horizontal force on the airplane must both also equal zero:

$$\sum F_x = F_T - F_D = 0 \qquad \sum F_y = F_L - F_g = 0$$

Therefore, thrust must equal drag in the horizontal direction, and lift must equal the attraction due to gravity in the vertical direction:

$$F_T = F_D \qquad\qquad F_L = F_g$$

The airplane will continue moving with a constant velocity (ie, a straight line with constant speed) until a change in one or more forces causes a nonzero net force on the airplane.

Newton's first law of motion implies that an object with a net force equal to zero remains at rest or moves with a constant velocity due to its inertia. Cases where the net force acting on an object is nonzero are addressed by **Newton's second law of motion**, which is discussed in the next section.

✓ Concept Check 1.8

The engines of an airplane provide a forward thrust equal to 10,000 N. When the airplane is cruising at an altitude of 30,000 ft with a constant velocity, what is the magnitude of the force of air drag opposing the motion of the plane?

Solution

Note: The appendix contains the answer.

1.3.02 Newton's Second Law

According to Newton's second law of motion, when the net force acting on an object is not equal to zero, the object changes velocity (ie, accelerates). For example, the hockey puck in Figure 1.25 is initially at rest on the ice. The constant force *F* applied to the puck causes it to start moving and increase speed. Unbalanced forces cause objects to change speed, direction, or both speed and direction.

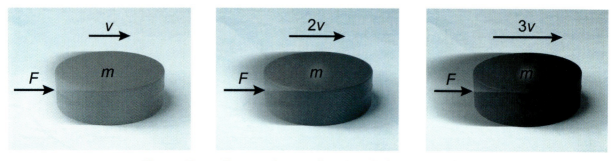

$F_{net} \neq 0$, so the puck accelerates (*v* increases)

Figure 1.25 A constant force causes a hockey puck to accelerate.

The resulting **acceleration** is directly proportional to the **net force** F_{net}. In other words, a car accelerating at rate *a* while the net force F_{net} acts on it accelerates at rate 2*a* when the net force of $2F_{net}$ acts on the same car.

However, since **mass** resists acceleration (because of inertia), the resulting acceleration is inversely proportional to the mass *m* of the object. An object with *twice* the mass as another accelerates at *half* the rate as another object given the same net force applied to both objects.

Consequently, the relationship between net force, mass, and acceleration is represented by:

$$\vec{a} = \frac{\vec{F}_{net}}{m}$$

or by rearranging into the more familiar form where F_{net} is the product of mass and acceleration:

$$\vec{F}_{net} = m\vec{a}$$

Please note that since acceleration and net force are **vector** quantities, they always act in the *same* direction.

As an example for calculating the acceleration of an object using Newton's second law of motion, Figure 1.26 shows two forces acting on a box in the same direction (rightward and leftward). The acceleration of each box is calculated by first determining the net force on the box F_{net} and then dividing that net force by the mass *m* of the box.

Chapter 1: Motion, Force, and Energy

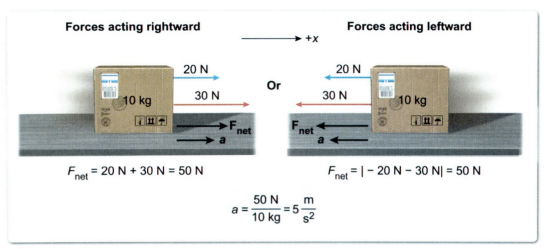

Figure 1.26 The acceleration of a box when forces act in the same direction, either leftward or rightward.

Assuming rightward is the positive direction, the net force F_{net} equals the sum of the two forces:

$$\vec{F}_{net} = (20\text{ N}) + (30\text{ N}) = 50\text{ N}$$

The magnitude of the acceleration of the box can be found by substituting F_{net} and mass m into Newton's second law equation:

$$|\vec{a}| = \frac{|\vec{F}_{net}|}{m} = \frac{(50\text{ N})}{(10\text{ kg})} = 5\ \frac{\text{m}}{\text{s}^2}$$

When the net force is rightward, the acceleration is also rightward, but when the net force is leftward, the acceleration is leftward.

Alternatively, in Figure 1.27, the forces on the box act in opposite directions.

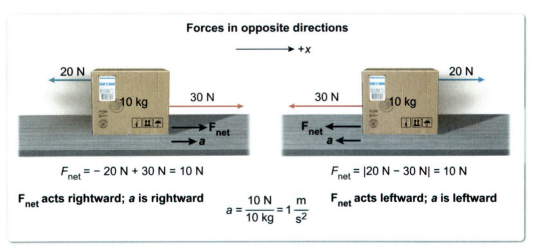

Figure 1.27 The acceleration of a box when forces act in opposite directions.

Assuming rightward is positive, the net force F_{net} equals the difference between the two forces:

$$\vec{F}_{net} = (30 \text{ N}) - (20 \text{ N}) = 10 \text{ N}$$

The magnitude of the acceleration of the box is found:

$$|\vec{a}| = \frac{|\vec{F}_{net}|}{m} = \frac{(10 \text{ N})}{(10 \text{ kg})} = 1 \frac{\text{m}}{\text{s}^2}$$

Alternatively, by knowing the values for the acceleration and mass of an object, the value of the net force acting on the object can be determined.

For example, the net force acting on a student (see Figure 1.28) with mass 50 kg accelerating to the right at 0.5 m/s² can be found using Newton's second law equation solved for F_{net}:

$$\vec{F}_{net} = m\vec{a} = (50 \text{ kg})\left(0.5 \frac{\text{m}}{\text{s}^2}\right) = 25 \text{ N}$$

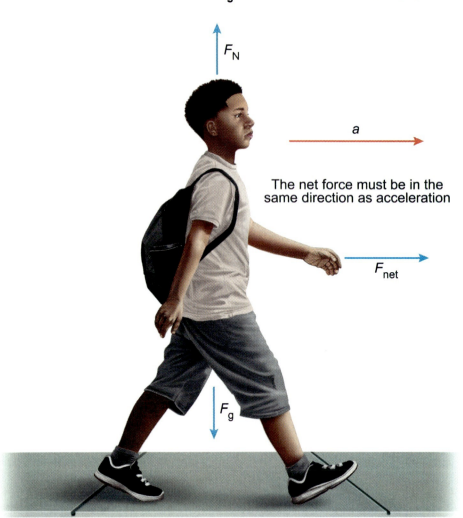

- $\sum \vec{F} = m\vec{a}$
- **a**   Acceleration
- F_N Normal force
- F_g Gravitational force (weight)

The net force must be in the same direction as acceleration

Figure 1.28 Student accelerating to the right due to a net force in the same direction.

Concept Check 1.9

Complete the following table by applying Newton's second law of motion:

	a (m/s²)	F_{net} (N)	m (kg)
Trial 1		30	2
Trial 2	10		5
Trial 3	7	21	

Solution

Note: The appendix contains the answer.

1.3.03 Newton's Third Law

Newton's third law of motion describes the relationship between objects in terms of **force pairs**. Every interaction between two objects produces two forces that are **equal** in magnitude and act in **opposite** directions on each object.

For example, when a bat strikes a baseball (Figure 1.29), the bat exerts a rightward force on the ball, causing the ball to accelerate to the right. At the same time, the ball exerts a leftward force of equal magnitude on the bat, causing the bat to accelerate to the left (ie, to slow down as it moves to the right).

Figure 1.29 The equal and opposite forces produced by the interaction between a bat and a baseball.

In general, if object A exerts a force on object B ($\vec{F}_{\text{A on B}}$), then object B exerts a force on object A ($\vec{F}_{\text{B on A}}$) with equal magnitude in the opposite direction:

$$\left|\vec{F}_{\text{A on B}}\right| = \left|\vec{F}_{\text{B on A}}\right|$$

$$\vec{F}_{\text{A on B}} = -\vec{F}_{\text{B on A}}$$

One way to identify Newton's third law force pairs (action/reaction) between two objects A and B is by the phrase shown in Figure 1.30.

If *A* acts on *B*, then *B* acts back on *A*

(Action)   (Reaction)

Figure 1.30 Newton's third law of motion involves an action and a reaction.

Hence, the action and reaction are *simultaneous*, *equal* in magnitude, and *opposite* in direction.

> ☑ **Concept Check 1.10**
>
> Identify the reaction forces to the following actions:
>
> a. Hammer strikes a nail
> b. Boxer punches a bag
> c. Earth pulls on the Moon
> d. Student pulls down on a rope anchored to the ceiling
> e. Student pulls up on a bathroom sink anchored to the floor
>
> **Solution**
>
> *Note: The appendix contains the answer.*

Although the force pairs associated with Newton's laws are equal and opposite, they do not have a canceling effect on one another. Two forces can only cancel if they act on the *same* object. Action and reaction force pairs act on *different* objects.

In Figure 1.31, a sliding can of beans collides with a stationary ball. Consequently, the can exerts a rightward force on the ball ($F_{\text{Can on ball}}$). As part of the interaction, the ball simultaneously exerts a force of equal magnitude leftward on the can ($F_{\text{Ball on can}}$). These two forces cause different changes in motion (ie, acceleration) for the can and the ball.

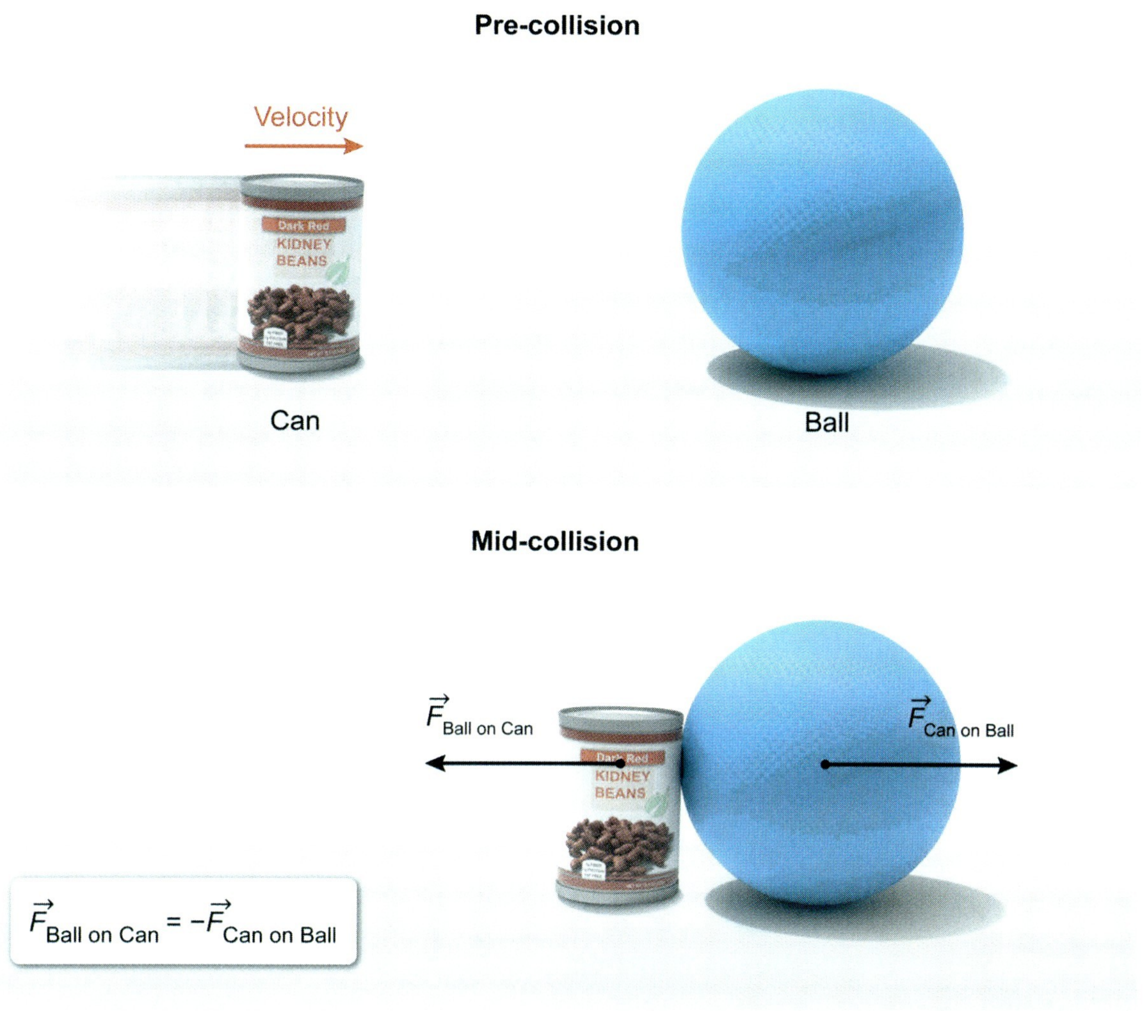

Figure 1.31 Force pairs during a collision between a can of beans and a ball.

Furthermore, two forces with equal magnitude acting in opposite directions are not necessarily an action/reaction force pair. For example, the reaction force to the downward pull of gravity on an object is not the upward push of the ground on that object. As seen in Figure 1.32, three force pairs exist for a student sitting in a chair. In one force pair, the Earth exerts a downward gravitational pull on the student $F_{\text{Earth on student}}$, and the student exerts an upward pull on Earth $F_{\text{Student on Earth}}$ with an equal magnitude.

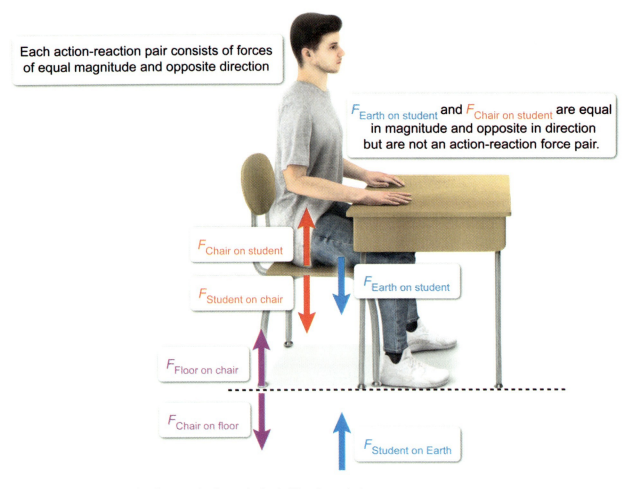

Figure 1.32 Action-reaction force pairs for a student sitting in a chair.

In a second interaction, the student exerts a downward contact force on the chair $F_{\text{Student on chair}}$, and the chair exerts a normal force of equal magnitude upward on the student $F_{\text{Chair on student}}$. Likewise, the third interaction is comprised of the equal magnitude downward contact force of the chair on the floor $F_{\text{Chair on floor}}$ and the upward normal force of the floor on the chair $F_{\text{Floor on chair}}$. The downward $F_{\text{Earth on student}}$ is equal in magnitude to the upward $F_{\text{Chair on student}}$, but the two forces do not comprise an action/reaction force pair. The reaction to the *downward* $F_{\text{Earth on student}}$ is the *upward* $F_{\text{Student on Earth}}$.

Newton's third law of motion aids in determining which forces act on objects. The resulting change in motion (ie, acceleration) or non-motion of those objects is determined by Newton's second and first laws of motion, respectively. Therefore, together, Newton's three laws of motion explain the motion of all objects.

 Concept Check 1.11

Two ice skaters initially at rest push off from each other and accelerate in opposite directions. Skaters *A* and *B* have masses of 60 kg and 40 kg, respectively. Compare the magnitudes of the accelerations of the two skaters.

Solution

Note: The appendix contains the answer.

1.3.04 Applications of Newton's Laws

Solving problems involving forces and motion requires identifying all the forces acting on an object, creating a **free-body diagram** with those forces, and applying **Newton's second law** of motion to the forces.

Forces can have a variety of origins. Some common forces include the gravitational force, the normal force, and tension. Other forces such as **friction** and the **spring force** are discussed later in the lesson.

Gravitational Force

The downward pull of Earth on an object is the gravitational force (ie, weight) and is often labeled F_g, as shown in Figure 1.33. The magnitude of the gravitational force can be determined as the product of mass m and the free-fall acceleration g (as mentioned in Concept 1.2.05, $g = 9.8 \text{ m/s}^2$):

$$F_g = mg$$

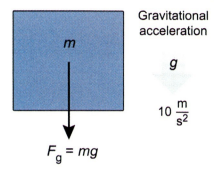

Figure 1.33 Weight is the downward gravitational force on a mass.

Normal Force

When two objects come into direct contact (ie, at their surfaces), normal forces F_N exist between them according to Newton's third law of motion (described in Concept 1.3.03). These forces act perpendicular to (ie, "normal" to) the surfaces of each object.

When an object is in contact with a horizontal surface (eg, floor, table, ground), a normal force acts upward on the object and downward on the surface (eg, Figure 1.34).

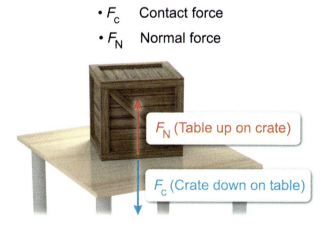

Figure 1.34 Normal forces acting on a crate and a table.

If no additional forces act on the object, the upward normal force F_N ("Table up on Crate") must be equal in magnitude to the downward gravitational force on the crate F_g:

$$F_N = F_g$$

If the surface is not horizontal, or an additional force has a vertical component, then the normal force is less than or greater than the gravitational force.

Tension Force

The force that a fixed rope, chain, or string exerts on an object is called the **tension** force F_T, which only acts along the length of the rope or chain. In the problems encountered on the exam, the tension force is constant along the length of the rope. In Figure 1.35, the gravitational force acts downward and the tension force acts upward, along the direction of the cable.

- F_T Tension force
- F_g Gravitational force (weight)

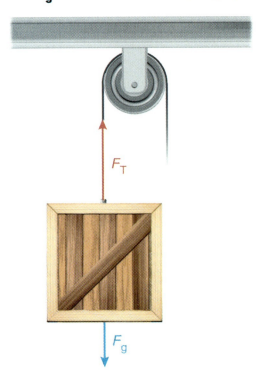

Figure 1.35 Tension force acting on a crate.

When a force makes an angle with the x- or y-axis, the force vector can be resolved into components along the axes. For example, Figure 1.36 illustrates the force of gravity, the normal force, and the tension force applied to a sled.

Since F_T makes an angle θ with the x-axis, its components along the x- and y-axes should be determined and used to solve problems:

$$F_{T,x} = F_T \cos\theta \qquad F_{T,y} = F_T \sin\theta$$

Chapter 1: Motion, Force, and Energy

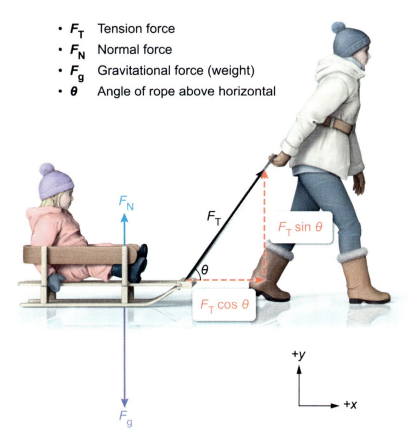

Figure 1.36 F_T makes an angle θ with the x-axis, so its components are calculated using sine and cosine.

Free-Body Diagrams

Solving problems often requires identifying all the forces acting on the object in question. A free-body diagram shows all forces acting on a single object in a frame of reference with meaningful labels to aid in identifying each force. The force diagram on the left in Figure 1.37 shows all the forces acting on a crate as it is pulled to the left by the tension in a rope and pushed to the right by a person.

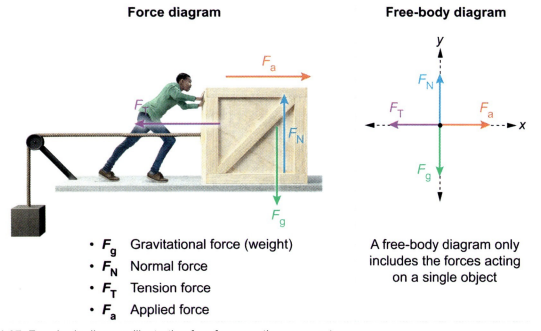

Figure 1.37 Free-body diagram illustrating four forces acting on a crate.

39

The forces acting on the crate include the downward gravitational force (ie, weight) F_g, the leftward tension force F_T, the rightward applied force F_a, and the upward normal force F_N. In the **free-body diagram** on the right, only the forces are shown for clarity. An xy-coordinate axis is applied to give the forces a frame of reference.

According to **Newton's second law of motion**, the net force acting on an object along each coordinate axis is equal to ma. In Figure 1.37, assuming the positive direction is up and to the right:

$$F_{net,x} = \sum F_x = F_a + F_T = ma_x$$

$$F_{net,y} = \sum F_y = F_N - F_g = ma_y$$

These equations can be used to solve for unknown quantities like acceleration or one of the forces acting on the object.

When a problem involves an object on an incline, such as the box on the truck in Figure 1.38, the downward gravitational force F_g makes an angle θ with the y-axis, where θ is the angle of the incline.

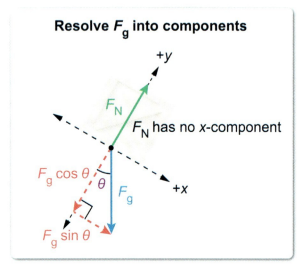
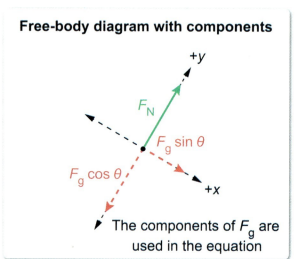

Figure 1.38 Free-body diagram for a box on the back of a truck.

$$F_{g,x} = F_g \sin\theta \qquad F_{g,y} = F_g \cos\theta$$

Once the components are determined, the free-body diagram is drawn, and the components are used in the Newton's second law of motion equations:

$$F_{net,x} = \sum F_x = F_g \sin\theta = ma_x$$

$$F_{net,y} = \sum F_y = F_N - F_g \cos\theta = ma_y$$

Concept Check 1.12

A ball is tossed straight up and allowed to fall straight down again. If air resistance is negligible, draw the free-body diagram for the ball when it is:

a. Halfway to its highest point on the way up.
b. At the highest point of its motion
c. Halfway from its highest point on the way down.

Solution

Note: The appendix contains the answer.

Concept Check 1.13

A person pulls a sled across the ground by an attached rope that makes an angle θ with the horizontal. The forces on the sled are the tension in the rope F_T, the gravitational force F_g, and the normal force F_N. Draw a free-body diagram showing the forces acting on the sled, use Newton's second law of motion to create an equation, and solve the equation for describe the change in the magnitude of the normal force F_N on the sled as the angle θ decreases.

Solution

Note: The appendix contains the answer.

1.3.05 Friction

Friction refers to the force that resists sliding between two surfaces that are in contact. Friction negates (partially or completely) any forces that promote sliding (ie, friction decreases acceleration). Intermolecular forces between the two surfaces contribute to **frictional forces** that oppose relative motion along the interface between two objects.

In **static equilibrium** (ie, motionless, with a net force equal to zero), a frictional force that resists the initiation of motion is called **static friction**. Static friction is oriented parallel to the contacting surfaces and arises in response to a force that attempts to move an object within the system.

For example, in Figure 1.39 a student exerts a rightward force on a box F_{push}, but the box remains at rest because a static frictional force between the box and the floor F_s acts leftward on the box with equal magnitude. As the student increases F_{push}, the magnitude of the static frictional force F_s also increases.

$$F_{push} = F_s$$

Student applies force to the right, but the box does not move...

- F_{push} Applied force
- F_s Static friction force
- F_N Normal force
- F_g Gravitational force (weight)

...because the force of static friction acts on the box to the left.

Figure 1.39 The force of static friction acts on the box to oppose its motion.

The maximum magnitude of static friction $F_{s,max}$ between two opposing surfaces is equal to the product of the normal force F_N and the **coefficient of static friction** μ_s, a unitless coefficient related to the physical characteristics of the contacting surfaces:

$$F_{s,max} = \mu_s \cdot F_N$$

If the force applied to an object does not exceed $F_{s,max}$, the object does not move. Conversely, if $F_{s,max}$ is exceeded, the object accelerates and is opposed by the force of **kinetic friction**.

Kinetic friction F_k is the frictional force between two objects sliding against each other, and it is the product of the coefficient of kinetic friction μ_k (which is specific to the two surfaces) and the normal force F_N:

$$F_k = \mu_k \cdot F_N$$

In Figure 1.40, the student exerts a force on the box F_{push} that exceeds $F_{s,max}$. As a result, the box accelerates to the right. The kinetic frictional force acts leftward on the box.

Chapter 1: Motion, Force, and Energy

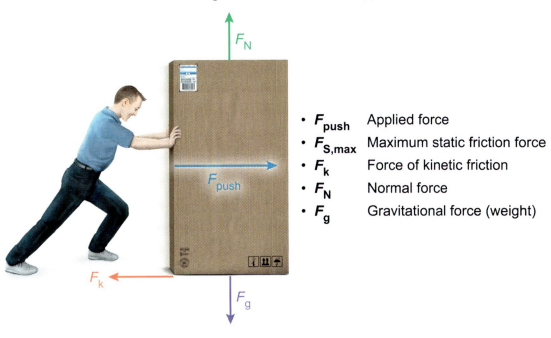

Figure 1.40 The force of kinetic friction opposes the motion of the box.

In general, kinetic friction does not depend on the *speed* of the objects or the *surface area* in contact. For most materials, the coefficient of static friction is greater than the coefficient of kinetic friction:

$$\mu_s > \mu_k$$

✓ Concept Check 1.14

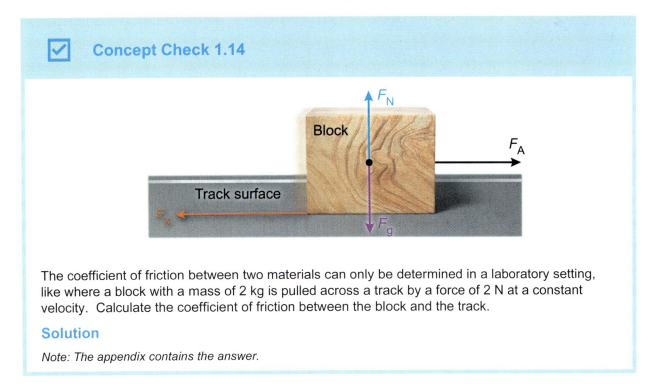

The coefficient of friction between two materials can only be determined in a laboratory setting, like where a block with a mass of 2 kg is pulled across a track by a force of 2 N at a constant velocity. Calculate the coefficient of friction between the block and the track.

Solution

Note: The appendix contains the answer.

> ☑ **Concept Check 1.15**
>
> A block slides along a horizontal surface where friction is not negligible. If the block has a mass of 5 kg and the coefficient of friction between the block and the surface is 0.2, calculate the acceleration of the block.
>
> **Solution**
>
> Note: The appendix contains the answer.

1.3.06 Hooke's Law

An **elastic force** F_{el} is generated by a spring or other elastic object when displaced from its **equilibrium position**, which is the natural position of the elastic object (ie, not stretched or compressed). The elastic force is directed opposite to the displacement and acts to restore the object to its equilibrium (ie, relaxed) position, as shown in Figure 1.41. Hence, it is also referred to as the restoring force.

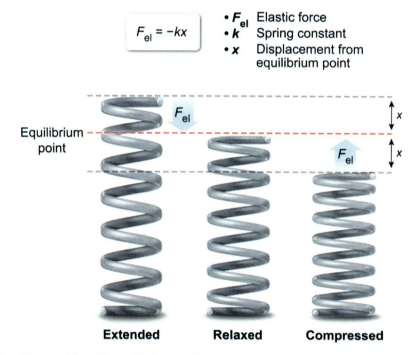

Figure 1.41 A spring displaced from its equilibrium position.

According to Hooke's law, F_{el} is a function of the spring **displacement** x and the elastic constant k for an **ideal spring**. This relationship is expressed mathematically as:

$$F_{el} = -kx$$

The value of k quantifies the relative elasticity of an object, which is attributable to its shape and the molecular structure of the material from which it is created. Although no material is perfectly elastic, Hooke's law is still useful in modeling the elastic forces generated by springs and other elastic objects.

For elastic objects that obey Hooke's law, the elastic force F_{el} is directly proportional to the distance x that the object is stretched from its natural length:

$$F_{el} \propto x$$

The relationship between elastic force and displacement is **linear** across the range of displacements, as shown in Figure 1.42.

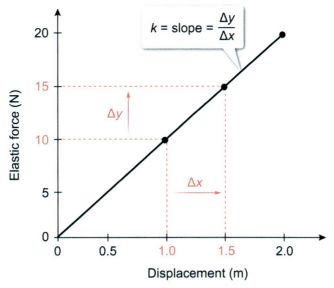

Figure 1.42 Determining k as the slope of a graph of Elastic force as a function of displacement.

The y-intercept is zero and the slope of the graph is equal to the value of the spring constant k:

$$k = \text{slope} = \frac{5\text{ N}}{0.5\text{ m}} = 10\ \frac{\text{N}}{\text{m}}$$

✓ Concept Check 1.16

A group of students measure the displacement of an elastic material while applying various forces and produce the graph of force vs. displacement shown here. Calculate the spring constant k of the material.

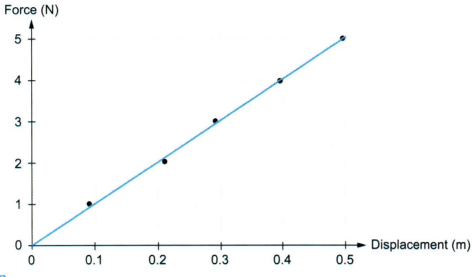

Solution

Note: The appendix contains the answer.

Chapter 1: Motion, Force, and Energy

Chapter 1: Motion, Force, and Energy

Lesson 1.4

Equilibrium

Introduction

This lesson begins by introducing the concept of the center of mass, which can be used to simplify the representation of a system of objects. When an external force acts on a rigid object or a system, the motion of the system's center of mass may change depending on the direction of the force and its point of application. Moreover, the force may also cause the system to begin rotating if a torque is generated. Thus, this lesson also explores the various variables and scenarios where a rotation will and will not occur.

Furthermore, the application of a force may not always result in a translational or rotational acceleration if other external forces are present. In this balanced state, an object or system can be either stationary (static) or in constant (dynamic) motion. The lesson concludes by discussing the difference between static equilibrium and dynamic equilibrium.

1.4.01 Center of Mass

A **system** is defined as a collection of objects, each with their own mass m_i and position r_i with respect to a reference point, as shown in Figure 1.43.

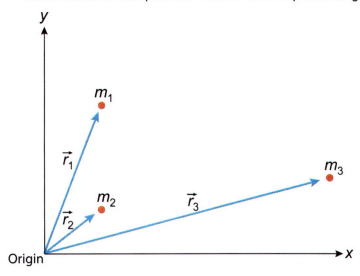

Figure 1.43 A system of objects with varying mass and position referenced on an *xy*-coordinate plane.

The mass of an entire system of objects can be characterized by a point in space known as the **center of mass (CM)**. The CM position r_{CM} represents the average of each object's displacement r_i from a reference point weighted by their mass m_i:

$$r_{CM} = \frac{m_1 r_1 + m_2 r_2 + m_3 r_3 + \cdots}{m_1 + m_2 + m_3 + \cdots}$$

Hence, the CM position represents the entire system as one combined mass concentrated at a single point in a position and is influenced more by the heavier objects in the system, as shown in Figure 1.44.

Chapter 1: Motion, Force, and Energy

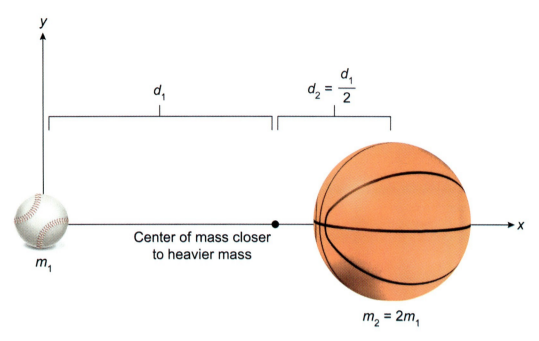

Figure 1.44 The CM of a system is closer to objects of greater mass.

For example, the CM position for an extended object, such as the human body (Figure 1.45), is closer to the areas of the body that contain more mass.

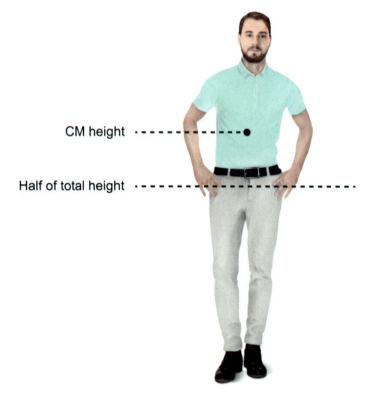

Figure 1.45 The CM position for a person is typically located above the vertical midpoint.

Furthermore, the CM position reduces the calculations that would consider every particle in the system or extended object to that of a single object. Thus, when a force acts on a system, it can be considered as acting on the system's CM.

Concept Check 1.17

A juggler throws four balls each of mass *m* upward into the air. What is the acceleration of the CM of the four-ball system? What is the net force acting at the CM? Leave your answer in terms of *m* and *g*.

Solution
Note: The appendix contains the answer.

 Concept Check 1.18

In the diagram below, the masses m_1, m_2, and m_3 are all equal. In which of the labeled quadrants is the CM position located?

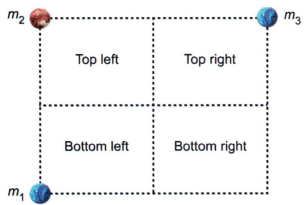

Solution
Note: The appendix contains the answer.

1.4.02 Torque

A **torque** represents the application of a force on an object at some radial distance away (known as the **lever arm**) from a pivot point, such as a hinge or axle. When a nonzero torque is applied to an object, the object accelerates in a rotational direction (ie, clockwise or counterclockwise). For this reason, torque is often referred to as the rotational version of a force.

The magnitude of a **torque** τ depends on the magnitude of the **applied force** F, the **lever arm distance** r, and the sine of the angle θ between F and r (Figure 1.46):

$$\tau = Fr \sin\theta$$

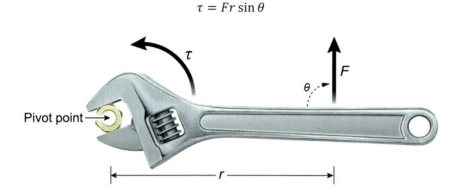

Figure 1.46 A perpendicular force F applied at a distance r away from the pivot point of a wrench generates a torque about that pivot point.

Consequently, any force applied in a direction directly toward or away from the rotational axis (or the center of mass for an unbound object) does not generate a torque (ie, the sine of 0° or 180°).

For a fixed force F, the magnitude of torque can be increased by increasing the lever arm distance or by varying θ toward 90°. Alternatively, decreasing r or varying θ toward 0° decreases the resulting torque.

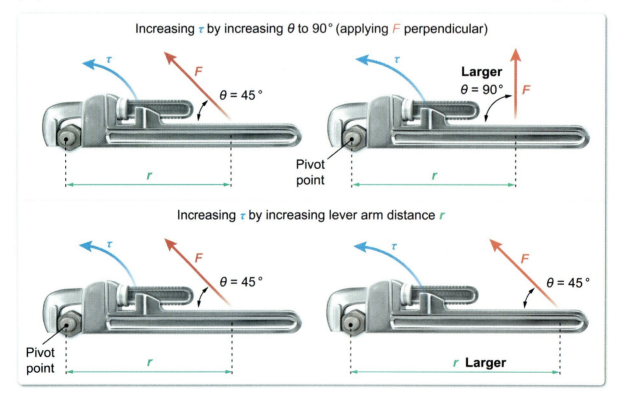

Figure 1.47 Increasing torque by increasing the lever arm or the sine of the angle between F and r.

Because the torque increases as a constant force is applied at greater distances from the pivot point (as shown in Figure 1.47), a mechanical advantage can be obtained.

☑ Concept Check 1.19

A plumber works with three identical valves and applies forces at either the edge or halfway to the edge on each attached lever. The forces have the magnitudes and angles indicated on each valve shown. Which valve experiences the greatest torque?

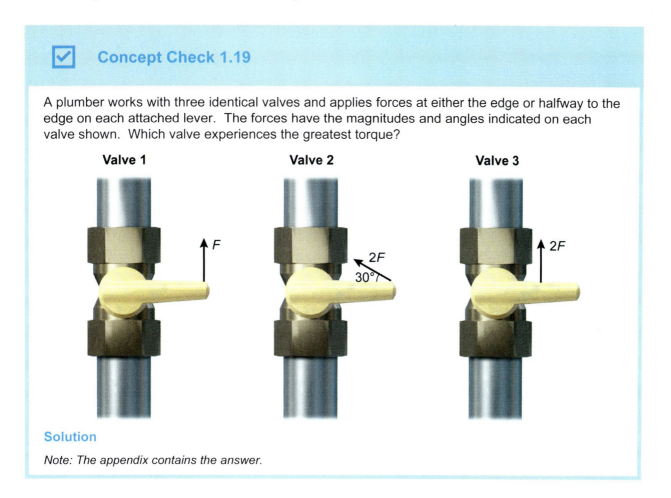

Solution

Note: The appendix contains the answer.

1.4.03 Static Equilibrium

As discussed in Concept 1.3.02, Newton's second law of motion implies that objects or systems accelerate linearly when a nonzero **net force** is present. Moreover, Newton's second law also applies to rotation, implying that an object **accelerates** angularly when a nonzero net torque is present.

However, when the sum of each force F_i and the sum of each torque τ_i acting on an object are equal to zero and the object is *not* moving (ie, linear velocity v and **angular velocity** ω both equal zero), the object is said to be in a state of **static equilibrium**:

$$\sum F_i = 0 \Rightarrow v = 0$$

$$\sum \tau_i = 0 \Rightarrow \omega = 0$$

For example, a stationary object located in two-dimensional space will not accelerate when the sum of all **translational forces** along both the x- and y-axes are equal to zero. For example, in Figure 1.48, a hanging block is pushed against a wall such that a vertical frictional force is generated to help counteract its weight. These forces, along each axis, can be organized by using a free-body diagram, where the forces acting on the object are organized and broken down into x- and y-vector components.

$$\sum x_{\text{forces}} = 0 = (+x \text{ forces}) + (-x \text{ forces})$$
$$\sum y_{\text{forces}} = 0 = (+y \text{ forces}) + (-y \text{ forces})$$

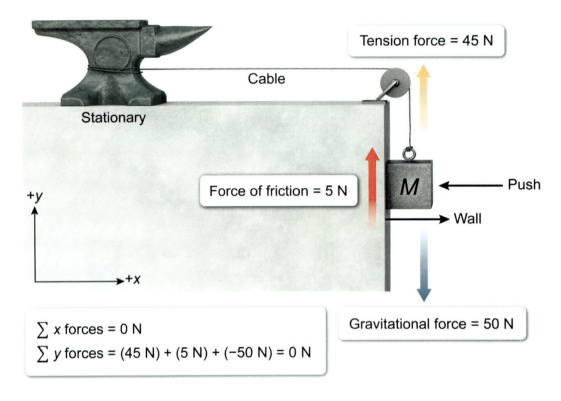

Figure 1.48 A frictional force and a tension force suspend a block in static equilibrium against the gravitational force (ie, weight).

Static equilibrium also occurs in a situation where two children are balanced on a stationary seesaw, which means that the net torque about the fulcrum must equal zero (Figure 1.49).

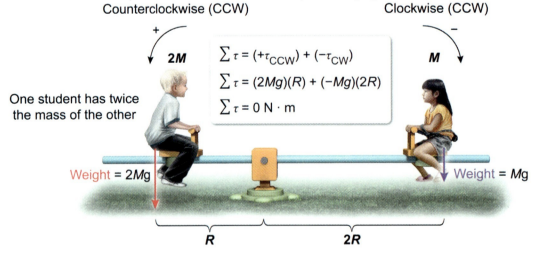

Figure 1.49 Two children balanced on a seesaw in static equilibrium.

Chapter 1: Motion, Force, and Energy

 Concept Check 1.20

A static friction force holds a box stationary on a surface inclined at an angle θ with respect to the horizontal. Draw a free-body diagram for the box and determine the net force equations in the x- and y-directions.

Solution

Note: The appendix contains the answer.

 Concept Check 1.21

A golfer can balance his golf club with his finger under a balance point A.

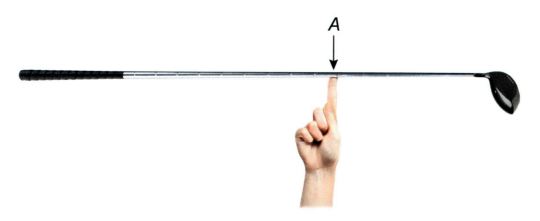

The clubhead is then removed and replaced with a clubhead of greater mass. How does the balance point change after the new clubhead is attached?

Solution

Note: The appendix contains the answer.

1.4.04 Dynamic Equilibrium

In the previous section, the condition of **static equilibrium** was shown to exist when the net force and the net torque on an object are equal to zero, and the object is at rest.

However, if the object is moving while the **net force** and **net torque** are zero, the object is instead in a state of **dynamic equilibrium**. Furthermore, from Newton's second law of motion, a net force and torque of zero implies that the object is not accelerating linearly or angularly, and must be moving or rotating at a **constant velocity**:

$$\sum F_i = 0 \Rightarrow v = \text{constant}$$

$$\sum \tau_i = 0 \Rightarrow \omega = \text{constant}$$

For example, an object moving at a constant speed can still experience a net force of zero because the sum of all **translational forces** along both the x- and y-axes equals zero at some point after the object was initially set into motion (Figure 1.50). By using a free-body diagram, where the forces acting on an object can be organized, each force can be resolved into x- and y-vector components and analyzed.

$$\sum x \text{ forces} = (+x \text{ forces}) + (-x \text{ forces}) = 0$$

$$\sum y \text{ forces} = (+y \text{ forces}) + (-y \text{ forces}) = 0$$

- F_f Frictional force
- $F_\parallel$ Component of weight parallel to incline
- M Mass
- g Acceleration of gravity
- θ Angle of incline
- $F_\perp$ Component of weight perpendicular to incline
- F_N Normal force

No net force

$$F_f = F_\parallel = Mg \cdot \sin\theta$$

$$F_\perp = F_N$$

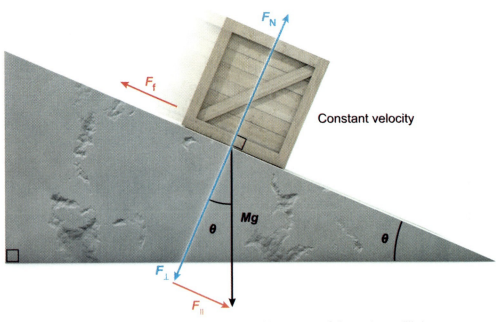

Figure 1.50 A box slides down an incline at a constant speed in a state of dynamic equilibrium.

In rotational motion, a state of **dynamic equilibrium** requires the torques (ie, rotational forces) to sum to zero while the object or system rotates at a constant speed. For example, Figure 1.51 shows a man applying a counterclockwise (CCW) torque to keep a platform rotating at a constant speed against the clockwise (CW) torque provided by internal frictional forces of the platform. To maintain dynamic equilibrium, the net torque acting on the platform must equal zero as it rotates.

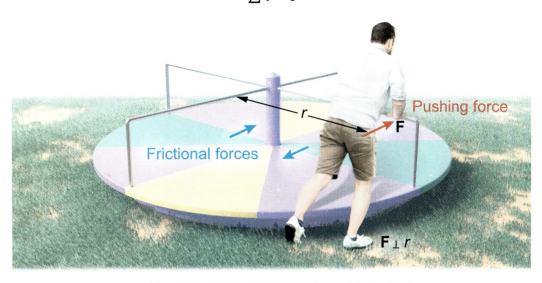

$$\sum \tau = (+\tau_{CCW}) + (-\tau_{CW})$$

$$\sum \tau = (\tau_{Pushing}) + (-\tau_{Friction}) = 0 \text{ N} \cdot \text{m}$$

Figure 1.51 A man pushes a platform to rotate it at a constant speed against internal forces.

Chapter 1: Motion, Force, and Energy

Lesson 1.5
Work and Energy

Introduction

Previous lessons describe how the forces applied to an object govern its motion, including the object's position, velocity, and acceleration. This lesson focuses on how these forces do work on an object, imparting different kinds of energy, and how the object's energy can be transformed from one type of energy to other types of energy.

1.5.01 Concept of Work

Mechanical work W refers to the energy transferred to or from an object when a force F is applied over a displacement d (see Figure 1.52). W is a scalar quantity, unlike F and d, which are vectors with magnitudes and directions.

Force is applied to the block over a displacement

Figure 1.52 Work is done on an object when applying a force over a displacement.

However, W does depend on the angle θ between F and d, as shown in Figure 1.53. In general, W is equal to the product of F, d, and the cosine of θ:

$$W = Fd \cos \theta$$

When F and d are in the same direction (ie, $\theta = 0°$), W is positive; when F and d are in opposite directions ($\theta = 180°$), W is negative. Furthermore, when F and d are perpendicular to each other ($\theta = 90°$), W is zero.

For specific situations where $\theta = 0°$, $\cos \theta$ is equal to 1 and W is calculated as simply the product of F and d:

$$W = F \cdot d$$

From this equation, the units of W are the product of newtons (N) and meters (m). The SI unit for W is the joule (J), where 1 J equals 1 N exerted over 1 m, or 1 J = 1 N·m.

Work (W) = Force (F) · Displacement (d) · cos θ

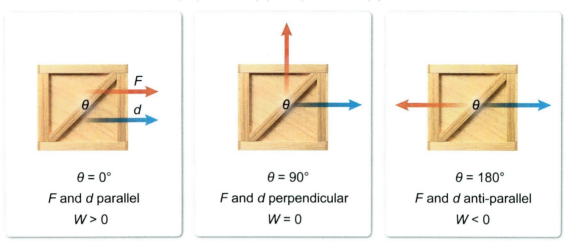

Figure 1.53 Work depends on the force, the displacement, and the angle between the directions of force and displacement.

The equation for work can be adapted to several different situations, such as an object pushed along a horizontal surface, an object sliding down an incline, or an object raised or lowered in a gravitational field.

For example, applying a horizontal force to push an object along a horizontal surface requires work to overcome the force of kinetic friction (Figure 1.54).

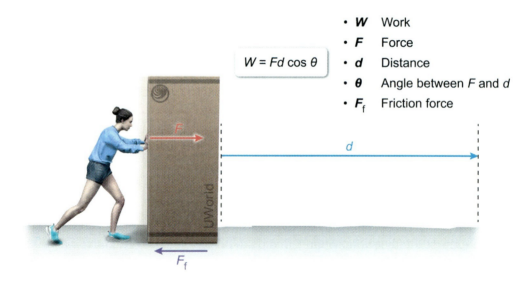

Figure 1.54 Work required to push an object along a horizontal surface to overcome force of kinetic friction.

Because both F and d are horizontal, θ = 0° and W equals the product of F, d, and the cosine of the angle θ between the vectors F and d:

$$W = Fd \cos 0° = Fd(1.0) = Fd$$

If a force F of 25 N is applied over a distance d of 10 m:

$$W = (25 \text{ N})(10 \text{ m}) = 250 \text{ J}$$

When a nonconstant horizontal force acts on an object, such as pushing a box along a horizontal surface, a plot of force versus displacement, as shown in Figure 1.55, can be used to calculate the work done.

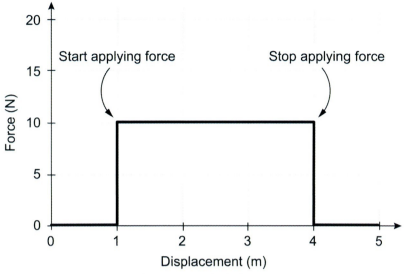

Figure 1.55 Plot of force versus displacement for a box pushed along a horizontal surface.

Because F and d are both horizontal, the angle θ between F and d equals 0°, and again W equals the product of F and d:

$$W = Fd \cos 0° = Fd$$

The plot shows that $F = 0$ N from $d = 0$ m to 1 m and $d = 4$ m to 5 m, resulting in $W = 0$ J during these intervals. However, $F = 10$ N from $d = 1$ m to 4 m, resulting in work done on the box:

$$W = (10 \text{ N})(4 \text{ m} - 1 \text{ m}) = (10 \text{ N})(3 \text{ m}) = 30 \text{ J}$$

Alternatively, W can be calculated by using the area under the curve to determine the product of F and d. The area under the curve equals the area of a rectangle A with a base b of 3 m and a height h of 10 N:

$$A = (3 \text{ m})(10 \text{ N}) = 30 \text{ J}$$

Consequently, W equals the product of A and $\cos 0°$:

$$W = A \cos \theta = (30 \text{ J})(1.0) = 30 \text{ J}$$

✓ Concept Check 1.22

A worker attempts to slide a heavy box along a rough floor but is unable to overcome the force of static friction acting on the box.

a. How much work is done on the box by the worker's pushing force?

b. How much work is done on the box by the static friction force?

Solution

Note: The appendix contains the answer.

1.5.02 Kinetic Energy and the Work-Energy Theorem

This concept focuses on a specific type of **energy**, kinetic energy, and its relationship to the work done on an object. An object can possess energy based on gravitational forces acting on it, referred to as potential energy PE, or based on its motion, referred to as kinetic energy KE. The KE of an object is equal to half the product of its mass m and its velocity v squared:

$$KE = \frac{1}{2}mv^2$$

From the equation above, the units of KE are the product of mass (kg) and velocity (m/s) squared. The SI unit for energy, including KE, is the joule (J).

Both kinetic energy and work have units of joules because work represents the transfer of energy to or from an object. The **work-energy theorem** states that the work W a force does on an object is equal to the change in the object's kinetic energy ΔKE:

$$W = \Delta KE$$

Negative W done on an object decreases its kinetic energy whereas positive W increases its kinetic energy. A person does positive W when kicking a ball because the KE of the ball increases, but a person does negative W when catching a ball because the KE decreases (see Figure 1.56).

Catching a ball
KE of ball decreases
W < 0 J

Kicking a ball
KE of ball increases
W > 0 J

Figure 1.56 Positive W is done when kicking a ball, and negative W is done when catching it.

The change in kinetic energy ΔKE is equal to the difference between the final kinetic energy KE_f and the initial kinetic energy KE_i:

$$W = \Delta KE = KE_f - KE_i$$

Typically, ΔKE results from a change in the object's velocity v because the object's mass is constant. In terms of the final velocity v_f and initial velocity v_i:

$$\Delta KE = \frac{1}{2}m(v_f^2 - v_i^2)$$

In agreement with Figure 1.56, this equation implies that positive W done on an object increases its velocity, and negative W decreases its velocity:

$$W = \frac{1}{2}m(v_f^2 - v_i^2)$$

As described in Concept 1.5.01, *W* is equal to the product of the applied force *F*, the displacement *d*, and the cosine of the angle θ between *F* and *d*:

$$W = Fd \cos \theta$$

Combining the equations for *W* and Δ*KE* yields:

$$Fd \cos \theta = \frac{1}{2} m(v_f^2 - v_i^2)$$

Many situations involve a constant *F*, including calculating the work done by gravity or an object's change in velocity. Given a change in kinetic energy for an object due to a constant *F*, the *F* or *d* required can be calculated.

However, the work done by a variable *F* can also be calculated. For example, the *F* acting on a 2 kg object at different positions as it slides along a horizontal surface is plotted in Figure 1.57.

If v_i equals 5 m/s at *d* = 0 m, v_f at *d* = 5 m can be determined by calculating *W* as the area under the curve on a plot of *F* versus *d*, as demonstrated in the final example in Concept 1.5.01.

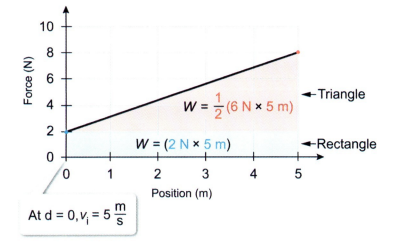

Figure 1.57 Variable force applied to an object as it slides along a horizontal surface.

The area includes a rectangle (bottom region) 2 N high and 5 m wide, and a triangle (top region) 6 N high and 5 m wide:

$$W = F \cdot d = (2 \text{ N} \times 5 \text{ m}) + \frac{1}{2}(6 \text{ N} \times 5 \text{ m})$$

$$W = 10 \text{ N} \cdot \text{m} + 15 \text{ N} \cdot \text{m} = 25 \text{ J}$$

Substituting the known values into the equation for the work-energy theorem gives:

$$25 \text{ J} = \frac{1}{2}(2 \text{ kg})\left(v_f^2 - \left(5\,\frac{\text{m}}{\text{s}}\right)^2\right)$$

$$\frac{(25)(2)}{2} = (v_f^2 - 25)$$

$$25 = v_f^2 - 25$$

$$v_f^2 = 25 + 25 = 50$$

$$v_f = \sqrt{50} \approx 7\,\frac{\text{m}}{\text{s}}$$

✓ Concept Check 1.23

A 100 kg box is at rest on a frictionless horizontal floor. A horizontal force of 20 N is applied to the box over a displacement of 10 m. What is the final velocity of the box?

Solution

Note: The appendix contains the answer.

1.5.03 Potential Energy of Systems

An object possesses **energy** based on its position in a system, referred to as its potential energy *PE*. For example, the *PE* of a system of objects near Earth's surface depends on the height *h* of the objects above a reference level, usually chosen to be the ground. For a single object, the *PE* of the object-Earth system is equal to the product of the object's mass *m*, the gravitational acceleration *g*, and *h*:

$$PE = mgh$$

In this formula, the units of *PE* are the product of mass (kg), acceleration (m/s²), and height (m). The SI unit for potential energy is the joule (J), 1 J = 1 (kg·m²)/s², the same unit as work and kinetic energy.

When the vertical position of an object changes, gravity does work on the object. The work *W* done by gravity on an object depends only on the change in the object's height Δ*h*, not on the path taken between the two positions:

$$W = -mg\Delta h = -\Delta PE$$

Note that this equation implies that negative *W* is done by gravity. This is a consequence of *g* being a downward force, whereas a positive Δ*h* is defined as an upward displacement. In other words, gravity does positive *W* when Δ*h* is negative (ie, the object has a downward displacement) and negative *W* when Δ*h* is positive (ie, the object has an upward displacement).

Because the work done by gravity is independent of the path taken (ie, depends only on the initial and final positions), gravity is a conservative force. Figure 1.58 shows the work done by gravity on two hikers with the same masses traveling from the base of a mountain to the summit is identical even though each hiker takes a different path to reach the summit.

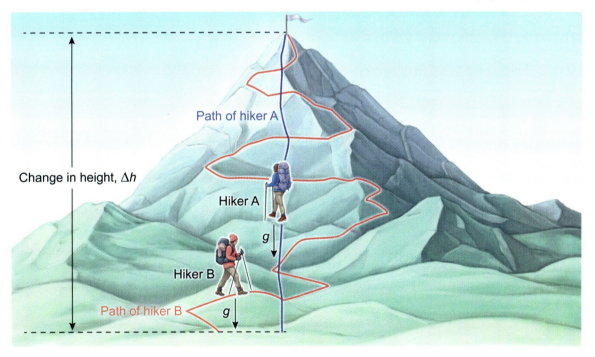

Figure 1.58 Work done by gravity is independent of the path taken, and is identical for hikers A and B.

Potential energy can also be stored in a mechanical system, such as a spring (Figure 1.59). The spring stores **elastic potential energy** PE_{el} when it is stretched or compressed from the equilibrium point (ie, its natural length). The PE_{el} stored in an **ideal spring** is equal to half the product of its spring constant k and its displacement x squared:

$$PE_{el} = \frac{1}{2}kx^2$$

Similar to the work done by gravity, the work done by an ideal spring is independent of the path taken, so the spring force is a conservative force. In other words, the PE_{el} of a spring compressed by x is independent of its initial condition, whether the spring was initially compressed, stretched, or at equilibrium.

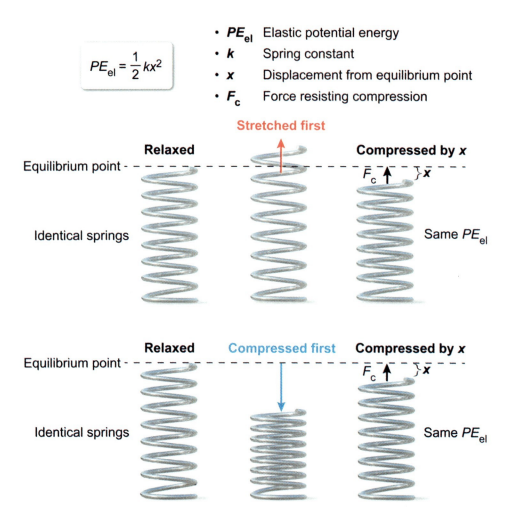

Figure 1.59 Elastic potential energy stored in a compressed spring.

Conversely, the work done by a **nonconservative force** like friction depends on the path taken and not just the initial and final positions.

Concept Check 1.24

A 1 kg crate is raised from a height of 0 m to a height of 4 m, and then lowered again to 0 m.

Work done by gravity to raise and lower crate

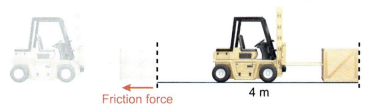

a. How much work did the gravitational force do on the crate?

The same crate is pushed along a horizontal surface that has a coefficient of kinetic friction of 0.3. The crate slides 4 m to the right, and then slides 4 m to the left.

Work done by kinetic friction on a crate

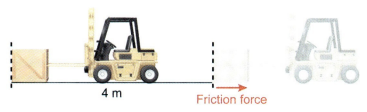

b. How much work did the friction force do on the crate?

Solution

Note: The appendix contains the answer.

1.5.04 Conservation of Total Energy

The total mechanical energy *E* of an object is the sum of the potential energy *PE* of the object-Earth system and the object's kinetic energy *KE*:

$$E = PE + KE$$

The *PE* for an object-Earth system is described in Concept 1.5.03 and is equal to the product of the object's mass *m*, the gravitational acceleration *g*, and its height *h*:

$$PE = mgh$$

Furthermore, Concept 1.5.02 showed an object's *KE* is equal to half the product of *m* and its velocity *v* squared:

$$KE = \frac{1}{2}mv^2$$

The law of conservation of energy states that *E* must remain constant when only conservative forces act on an object:

$$E = PE + KE = \text{constant}$$

This equation implies that *E* is transformed from *PE* to *KE* (and vice versa) while the total *E* remains constant. Conservation of energy is illustrated in Figure 1.60, where a baby is gently tossed into the air and then caught.

When the baby is first tossed into the air (Point A), their *PE* equals zero and *E* equals KE_A because their velocity v_A is at its maximum. When the baby is halfway to their maximum height (Point B), their velocity v_B has decreased and *E* equals the sum of PE_B and KE_B. At the maximum height (Point C), the baby's velocity v_C is zero and *E* equals PE_C. Note that the value of *E* is the same at each point because KE_A is completely converted into PE_C as the baby's height increases and their velocity decreases.

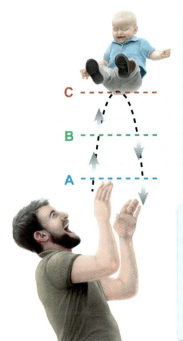

Figure 1.60 Conservation of energy states that the total energy remains constant but can transform from potential energy to kinetic energy, or vice versa.

In contrast, nonconservative forces, such as kinetic friction and air resistance, can convert some of an object's *E* into other forms (eg, heat, sound, light), so the sum of *PE* and *KE* is no longer constant, (ie, total *E* decreases).

The conservation of energy equation can be written in terms of the changes in potential energy Δ*PE* and kinetic energy Δ*KE*:

$$\Delta PE + \Delta KE = 0$$

Furthermore, these changes in energy can be calculated from the initial and final potential energy, *PE*$_i$ and *PE*$_f$, and the initial and final kinetic energy, *KE*$_i$ and *KE*$_f$:

$$(PE_f - PE_i) + (KE_f - KE_i) = 0$$

Therefore, the conservation of energy equation can be used to calculate the initial or final velocity given changes in *PE* or the initial and final positions given changes in *KE*. These variables are described in Figure 1.61, which shows a 0.2 kg ball initially at rest falling from a height of 6 m.

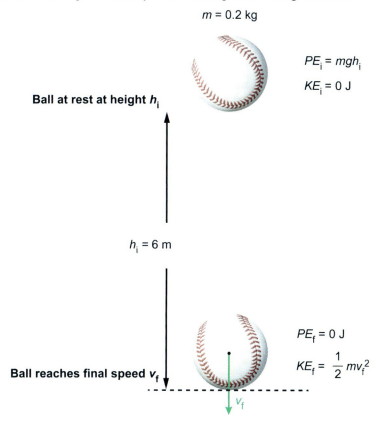

Figure 1.61 The change in kinetic energy of a 0.2 kg ball falling from a height of 6 m.

The *PE*$_i$ of the ball is equal to the product of *m*, *g*, and the initial height *h*$_i$:

$$PE_i = mgh_i = (0.2 \text{ kg})\left(10\,\frac{\text{m}}{\text{s}^2}\right)(6 \text{ m}) = 12 \text{ J}$$

Because the ball is initially at rest, its initial velocity *v*$_i$ equals zero. Thus, the *KE*$_i$ of the ball also equals zero:

$$KE_i = \frac{1}{2}mv_i^2 = 0 \text{ J}$$

Just before the ball hits the ground, its *PE*$_f$ equals zero because *h*$_f$ equals zero:

$$PE_f = mgh_f = 0 \text{ J}$$

This ΔPE for the ball must be balanced by an equal gain in the ball's ΔKE. Hence, its KE_f must equal its PE_i:

$$KE_f = PE_i = 12 \text{ J}$$

$$KE_f = \frac{1}{2}mv_f^2 = \frac{1}{2}(0.2 \text{ kg})v_f^2 = 12 \text{ J}$$

Solving for v_f yields:

$$v_f^2 = \frac{2}{0.2}(12) = 120$$

$$v_f = \sqrt{120} \approx 11 \frac{\text{m}}{\text{s}}$$

A similar approach can be used to calculate the compression of a spring when struck by a moving block (see Figure 1.62). The total E of the spring-block system must remain constant, so the kinetic energy of the block KE_B is completely transformed into elastic potential energy PE_{el} in the spring. Hence, the sum of the change in the spring's PE_{el} and the change in the block's KE_B equals zero:

$$\Delta PE_{el} + \Delta KE_B = 0$$

The initial PE_{el} equals zero because the spring is initially at equilibrium. The final PE_{el} equals half the product of its spring constant k and its maximum displacement x_{max} squared:

$$\Delta PE_{el} = \frac{1}{2}kx_{max}^2$$

The final KE_B equals zero because the block is at rest when the spring is fully compressed. Hence, the change in the block's KE_B equals:

$$\Delta KE_B = -KE_B$$

Combining these two equations yields:

$$\frac{1}{2}kx_{max}^2 + (-KE_B) = 0$$

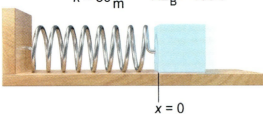

- **ΔPE_{el}** Change in elastic potential energy
- **ΔKE_B** Change in kinetic energy of block
- **k** Spring constant
- **x** Spring displacement
- **x_{MAX}** Maximum spring displacement

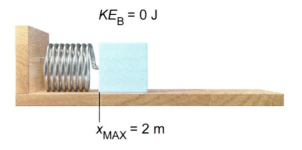

Figure 1.62 Compression of a spring struck by a moving block.

With $k = 50$ N/m and $KE_B = 100$ J, x_{max} of the spring is given by:

$$\frac{1}{2}kx_{max}^2 + (-KE_B) = \frac{1}{2}\left(50\,\frac{N}{m}\right)x_{max}^2 - 100\,J = 0$$

$$x_{max}^2 = \frac{2}{50}(100) = 4$$

$$x_{max} = \sqrt{4} = 2\,m$$

✓ Concept Check 1.25

A 0.5 kg ball is thrown straight up with an initial velocity of 10 m/s. What is the velocity of the ball when it reaches a height of 1 m?

Solution

Note: The appendix contains the answer.

1.5.05 Power

Power P is defined as the transfer of energy E per unit time t:

$$P = \frac{E}{t}$$

In mechanics, the transfer of energy is equal to the work W done by a force, so P can be calculated from W and t:

$$P = \frac{W}{t}$$

The SI unit for power is the watt (W), where 1 W equals 1 joule (J) of work done over 1 second (s) of time.

Concept 1.5.01 discussed how work can be applied to a variety of situations, including raising an object, pulling an object along a horizontal surface, and sliding an object down an incline. All these situations can also be described by the power required for a force to do work on an object. Work W is equal to the product of the constant force F applied and the displacement d over which it is applied:

$$W = Fd$$

Substituting this equation for W into the equation for P yields:

$$P = \frac{Fd}{t} = F\frac{d}{t}$$

Since the ratio of d and t in the above equation is equal to a constant velocity v, P can also be calculated as the product of F and v:

$$P = Fv$$

A sled with a kinetic friction force F_f of 100 N is pulled along a horizontal surface at a constant velocity v of 0.5 m/s (Figure 1.63).

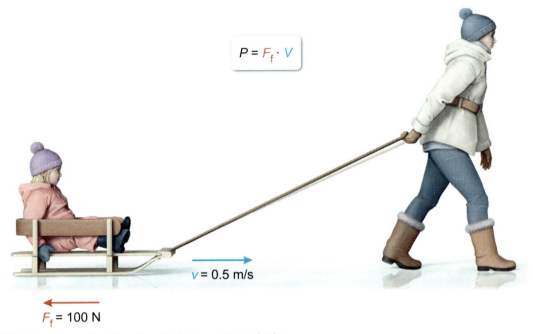

Figure 1.63 Power required to pull a sled at constant velocity.

Because the sled travels at a constant speed (ie, has zero acceleration), Newton's second law of motion requires the two horizontal forces acting in opposite directions on the sled, F_f and the horizontal component of the pulling force F_P, to be equal:

$$F_P = F_f$$

Therefore, the power P required to pull the sled equals the product of F_P and v:

$$P = Fv = (100 \text{ N})\left(0.5 \, \frac{\text{m}}{\text{s}}\right) = 50 \text{ W}$$

From Concepts 1.5.02 and 1.5.03, W is equal to the change in the kinetic energy ΔKE or the potential energy ΔPE of an object:

$$W = \Delta KE$$

$$W = \Delta PE$$

Hence, P equals the ratio of ΔKE or ΔPE and the time t that the force acts on the object:

$$P = \frac{\Delta KE}{t}$$

$$P = \frac{\Delta PE}{t}$$

For example, the power required for a horse to lift a 15 kg stone to a height of 1 m in 1 s can be calculated as shown in Figure 1.64. The ΔPE of the stone is equal to the product of the mass m, the gravitational acceleration g, and the change in height Δh:

$$\Delta PE = mg \cdot \Delta h$$

Substituting the values into this equation yields:

$$\Delta PE = (15 \text{ kg})\left(10 \, \frac{\text{m}}{\text{s}^2}\right)(1 \text{ m}) = 150 \text{ J}$$

Furthermore, this ΔPE occurs over a t of 1 s, resulting in P given by:

$$P = \frac{\Delta PE}{t} = \frac{150 \text{ J}}{1 \text{ s}} = 150 \text{ W}$$

$$P = \frac{W}{t} = \frac{mg \cdot \Delta h}{t}$$

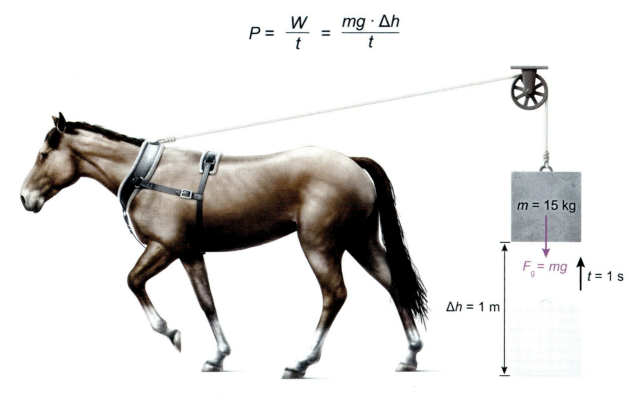

Figure 1.64 Power required to lift a 15 kg mass by 1 m in 1 s.

When the force acting on the object is not constant, the average power over a specific time is calculated from the average force or the change in the object's mechanical energy. The force acting on a box as it slides a displacement d of 20 m along a surface is plotted in Figure 1.65 and the average power over the time shown in the graph can be calculated.

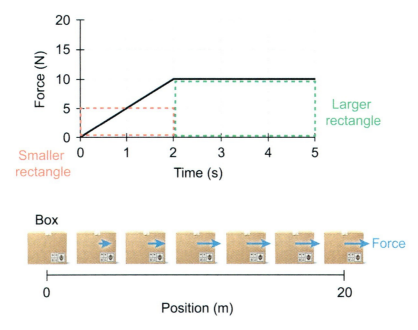

Figure 1.65 Plot of force acting on a box versus time as it slides 20 m along a horizontal surface.

The average force acting from 0 s to 2 s (the smaller rectangle) equals 5 N, and the force equals 10 N from 2 s to 5 s (the larger rectangle). Therefore, the average force F_{avg} over the full 5 s duration is:

$$F_{avg} = \frac{(5 \text{ N} \times 2 \text{ s}) + (10 \text{ N} \times 3 \text{ s})}{5 \text{ s}} = \frac{(10 \text{ N} \cdot \text{s}) + (30 \text{ N} \cdot \text{s})}{5 \text{ s}} = \frac{40 \text{ N}}{5} = 8 \text{ N}$$

The work W done equals the product of F_{avg} and d:

$$W = F_{avg} \cdot d = (8 \text{ N})(20 \text{ m}) = 160 \text{ J}$$

Hence, the average power P_{avg} equals the ratio of W and t:

$$P_{avg} = \frac{W}{t} = \frac{160 \text{ J}}{5 \text{ s}} = 32 \text{ W}$$

> ### Concept Check 1.26
>
> A 1,200 kg car accelerates from rest to 30 m/s in 15 s. What is the average power of the car's engine?
>
> **Solution**
>
> *Note: The appendix contains the answer.*

1.5.06 Mechanical Advantage and Simple Machines

Simple machines can be used to amplify forces through the phenomenon of mechanical advantage. Simple machines include the pulley, the lever, and the inclined plane. The **mechanical advantage** of these systems is equal to the ratio of the output force F_o to the input force F_i:

$$\text{Mechanical advantage} = \frac{F_o}{F_i}$$

Pulleys

There are two types of **pulleys**: a fixed pulley and a movable pulley (see Figure 1.66). The **fixed pulley** is connected to an immovable point, such as a wall or ceiling, and only changes the direction of the tension force in a rope, not the tension force itself. To lift a load, the pully must generate a F_o equal to the load's weight W, ie the product of the load's mass m and gravitational acceleration g:

$$F_o = W = mg$$

Because the pully does not alter the tension force on the rope, F_i is also equal to W:

$$F_i = W = mg$$

Hence, the mechanical advantage of a fixed pulley is equal to one:

$$\frac{F_o}{F_i} = \frac{mg}{mg} = 1$$

A **movable pulley** is connected to the load, so it moves when the load moves. When the load is attached to the pulley, the W is supported by both ends of the rope, decreasing the tension force on the rope by a factor of 2. As a result, the pully generates a F_o equal to W:

$$F_o = W = mg$$

Because the tension force on the rope is half of W, F_i is equal to half of W:

$$F_i = \frac{W}{2} = \frac{mg}{2}$$

Hence, the mechanical advantage of a movable pulley is equal to two:

$$\frac{F_o}{F_i} = \frac{mg}{\left(\frac{mg}{2}\right)} = 2$$

Mechanical advantage = $\frac{F_o}{F_i}$

- F_o Output force
- F_i Input force

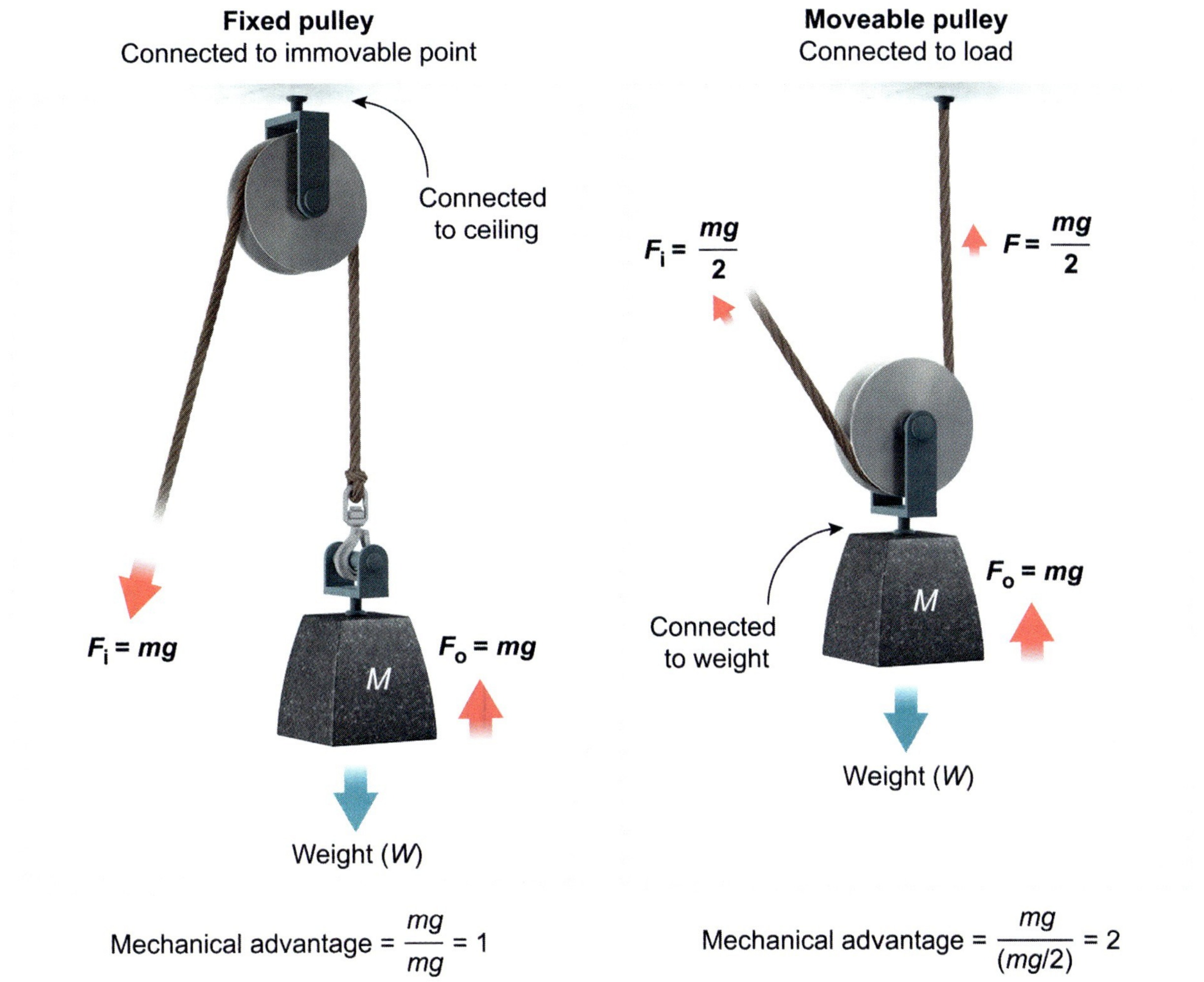

Fixed pulley
Connected to immovable point

Moveable pulley
Connected to load

Connected to ceiling

$F_i = \frac{mg}{2}$

$F = \frac{mg}{2}$

$F_i = mg$

$F_o = mg$

$F_o = mg$

Connected to weight

Weight (W)

Weight (W)

Mechanical advantage = $\frac{mg}{mg} = 1$

Mechanical advantage = $\frac{mg}{(mg/2)} = 2$

Figure 1.66 A fixed pulley with a mechanical advantage of 1 and a movable pulley with a mechanical advantage of 2.

Fixed and movable pulleys can be combined into a pulley system, as shown in Figure 1.67. A system of only fixed pulleys has a mechanical advantage of 1. A system of only moveable pulleys or a combination of fixed and movable pulleys has an overall mechanical advantage equal to the product of 2 and the number n_{MP} of moveable pulleys in the system:

$$\text{Mechanical advantage} = 2 \cdot n_{MP}$$

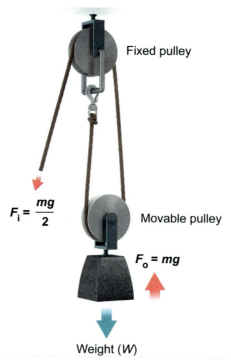

Figure 1.67 Pulley system consisting of one fixed pulley and one moveable pulley with a mechanical advantage of 2.

Levers

A lever can transfer a small force applied to one end of the lever into a larger force on the other end (Figure 1.68).

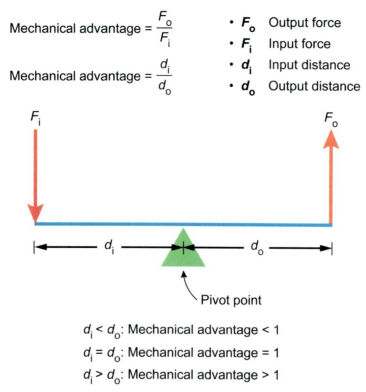

$$\text{Mechanical advantage} = \frac{F_o}{F_i}$$

$$\text{Mechanical advantage} = \frac{d_i}{d_o}$$

- F_o Output force
- F_i Input force
- d_i Input distance
- d_o Output distance

$d_i < d_o$: Mechanical advantage < 1
$d_i = d_o$: Mechanical advantage = 1
$d_i > d_o$: Mechanical advantage > 1

Figure 1.68 Input and output forces on a lever using a pivot point to generate a mechanical advantage.

The mechanical advantage achieved with a lever depends on the distances of the input force F_i and output force F_o relative to the pivot point of the lever. Applying F_i a distance d_i from the pivot point generates F_o a distance d_o on the opposite side of the pivot point according to the relationship:

$$\frac{F_o}{F_i} = \frac{d_i}{d_o}$$

Therefore, the mechanical advantage of a lever is equal to the ratio of d_i and d_o:

$$\text{Mechanical advantage} = \frac{d_i}{d_o}$$

The above equation shows that the mechanical advantage is 1 when d_i equals d_o, and a mechanical advantage greater than 1 is achieved when d_i is greater than d_o.

Inclined planes

Another simple machine that can provide a mechanical advantage is the **inclined plane**. An inclined plane is assumed to be frictionless and can be used to lift an object to a height by applying a force that is less than the object's weight.

Figure 1.69 shows an object with mass m on an inclined plane experiences a vertical gravitational force F_g equal to the product of m and the gravitational acceleration g (ie, the object's weight):

$$F_g = mg$$

Recall from Concept 1.3.04 that F_g can be separated into a force $F_\parallel$ parallel to the inclined plane and a force $F_\perp$ perpendicular to the inclined plane. Both $F_\perp$ and the normal force F_N are equal to the product of m, g, and the cosine of the inclined plane's angle θ:

$$F_\perp = F_N = mg \cos \theta$$

$F_\parallel$ represents the force acting to slide the object down the incline, and is equal to the product of m, g, and $\sin \theta$:

$$F_\parallel = mg \sin \theta$$

- F_g Weight
- m Mass
- g Gravitational acceleration
- $F_\perp$ Perpendicular force
- F_N Normal force
- θ Incline angle
- $F_\parallel$ Parallel force

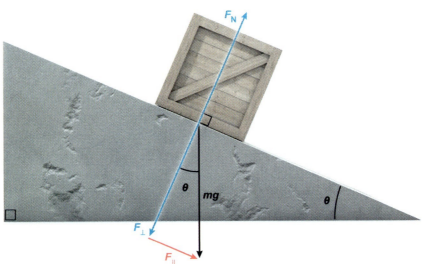

Figure 1.69 Forces acting on an object on an inclined plane.

Sliding the object up the ramp requires applying an input force F_i equal to $F_\parallel$: (Figure 1.70)

$$F_i = mg \sin \theta$$

The output force F_o for the ramp is equal to the force required to raise the object without a ramp (ie, the object's weight):

$$F_o = mg$$

Therefore, the mechanical advantage for an inclined plane is equal to the ratio of F_o and F_i:

$$\text{Mechanical advantage} = \frac{F_o}{F_i} = \frac{mg}{mg \sin \theta} = \frac{1}{\sin \theta}$$

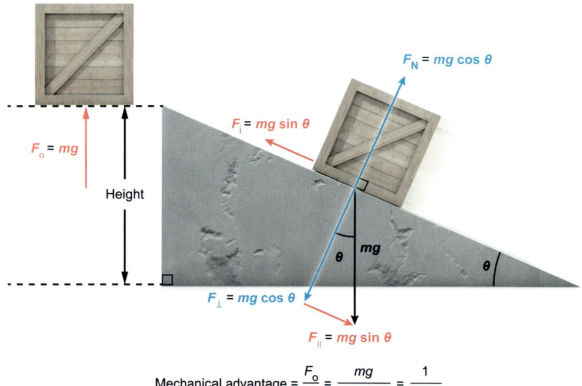

Mechanical advantage = $\frac{F_o}{F_i}$
- F_o Output force
- F_i Input force

$$\text{Mechanical advantage} = \frac{F_o}{F_i} = \frac{mg}{mg \sin \theta} = \frac{1}{\sin \theta}$$

$\theta = 5°$: Mechanical advantage ≈ 12
$\theta = 30°$: Mechanical advantage = 2
$\theta = 45°$: Mechanical advantage ≈ 1.4
$\theta = 60°$: Mechanical advantage ≈ 1.2

Figure 1.70 Mechanical advantage of the inclined plane.

Recall that sin θ represents the ratio of the opposite side of a right triangle to the hypotenuse. Hence, the inverse of the sine function is the ratio of the hypotenuse to the opposite side. For an inclined plane, this ratio is equivalent to the length of the inclined plane over its height:

$$\text{Mechanical advantage} = \frac{1}{\sin \theta} = \frac{\text{Length of inclined plane}}{\text{Height of inclined plane}}$$

Work done by simple machines

Although pulleys, levers and inclined planes can provide a mechanical advantage, the overall work done on the object does not change. As discussed in Concept 1.5.01, work W is equal to the product of the force F and the distance d over which it is applied:

$$W = Fd$$

Although a mechanical advantage reduces the force required to lift an object, it also increases the distance required such that the same work is done on the object with or without the mechanical advantage. When using a pulley, this work requires pulling the rope a longer distance than the height the object is raised. Similarly, the input end of a lever must move a greater distance than the output end, and the length of an inclined plane increases as the mechanical advantage increases.

 Concept Check 1.27

For an inclined plane with an angle of 0.1°:

a. What is the mechanical advantage of the inclined plane?

b. How long must this inclined plane be to lift an object to a height of 2 m?

Solution

Note: The appendix contains the answer.

END-OF-UNIT MCAT PRACTICE

Congratulations on completing **Unit 1: Mechanics and Energy**.

Now you are ready to dive into MCAT-level practice tests. At UWorld, we believe students will be fully prepared to ace the MCAT when they practice with high-quality questions in a realistic testing environment.

The UWorld Qbank will test you on questions that are fully representative of the AAMC MCAT syllabus. In addition, our MCAT-like questions are accompanied by in-depth explanations with exceptional visual aids that will help you better retain difficult MCAT concepts.

TO START YOUR MCAT PRACTICE, PROCEED AS FOLLOWS:

1) Sign up to purchase the UWorld MCAT Qbank
   IMPORTANT: You already have access if you purchased a bundled subscription.
2) Log in to your UWorld MCAT account
3) Access the MCAT Qbank section
4) Select this unit in the Qbank
5) Create a custom practice test

Unit 2 Fluids

Chapter 2 Fluid Dynamics

2.1 Hydrostatics

    2.1.01   Density and Specific Gravity
    2.1.02   Hydrostatic Pressure
    2.1.03   Buoyancy and Archimedes' Principle
    2.1.04   Surface Tension and Capillary Action

2.2 Fluids in Motion

    2.2.01   Ideal vs Nonideal Flow
    2.2.02   Continuity Equation
    2.2.03   Bernoulli's Equation
    2.2.04   Poiseuille Viscous Flow and Turbulence
    2.2.05   Applications to Blood Circulation

Chapter 2: Fluid Dynamics

Lesson 2.1
Hydrostatics

Introduction

Solids are substances in which the atoms or molecules are held permanently in place and resist forces in any direction. Materials in which the bonds between atoms or molecules are more easily broken due to their greater internal energy are known as fluids. Fluids are not arranged with any regular structure, and therefore can be deformed by stresses and shear forces. Both gases and liquids can behave as fluids. This lesson discusses the properties of fluids at rest and Lesson 2.2 describes the physics of fluids in motion.

2.1.01 Density and Specific Gravity

The **density** ρ of a substance is defined as the ratio of a given sample's **mass** m and the **volume** V that it occupies:

$$\rho = \frac{m}{V}$$

Hence, the SI unit combination for density is kg/m³. For instance, consider a sample solution that has a volume equal to 150 μL and mass of 100 mg, as shown in Figure 2.1.

$\rho = \dfrac{m}{V}$

- ρ Density
- m Mass
- V Volume

$m = 100$ mg, $V = 150$ μL

$$\rho = \frac{100 \text{ mg}}{150 \text{ μL}}$$

Sample

Figure 2.1 Calculating the density of a substance.

The density of the sample can be found by substituting these values into the above equation, yielding:

$$\rho = \frac{100 \text{ mg}}{150 \text{ μL}}$$

Converting the units of mg to kg and μL to L gives:

$$\rho = \frac{(100 \text{ mg})\left(\frac{10^{-3} \text{ g}}{1 \text{ mg}}\right)\left(\frac{10^{-3} \text{ kg}}{1 \text{ g}}\right)}{(150 \text{ μL})\left(\frac{10^{-6} \text{ L}}{1 \text{ μL}}\right)}$$

$$\rho = 0.67 \, \frac{\text{kg}}{\text{L}}$$

However, the standard SI units for ρ are kg/m³. Because 1 L = 1,000 cm³, there are 10^{-3} L for each 1 cm³. From the relationship 1 m = 100 cm, cubing both sides gives 1 m³ = (100 cm)³ = 10^6 cm³. Therefore, converting ρ to standard SI units yields:

$$\rho = \left(0.67 \; \frac{\cancel{\text{kg}}}{\cancel{\text{L}}}\right)\left(\frac{10^{-3} \; \cancel{\text{L}}}{1 \; \cancel{\text{cm}^3}}\right)\left(\frac{10^6 \; \cancel{\text{cm}^3}}{1 \; \text{m}^3}\right)$$

$$\rho = 670 \; \frac{\text{kg}}{\text{m}^3}$$

If the density and volume of the fluid are known, then the mass of the sample can be determined by rearranging the above equation and calculating the product of ρ and V:

$$m = \rho V$$

For the exam, it is important to memorize the density of water (ρ_{water} = 1 kg/L) and be able to convert it into equivalent units (eg, 1 g/cm³).

 Concept Check 2.1

A spherical hot air balloon has a diameter of 10 m and is filled with air having a mass of 500 kg. What is the density ρ of the air contained within the balloon in g/L?

Solution

Note: The appendix contains the answer.

The **specific gravity** SG of a liquid or solid is a measure of the relative density of two samples and is useful in determining concentration and chemical composition. A sample's SG is determined as the ratio of the density of the substance ρ_{sub} to the density of water (ρ_{water} = 1 kg/L):

$$SG = \frac{\rho_{sub}}{\rho_{water}}$$

 Concept Check 2.2

The density of a woman's bones at age 40 is about 1.5 g/cm³. By age 80, her total bone mass has decreased by 1 kg. Assuming the volume of bone stays the same at 4,000 cm³, what is her total bone mass, density, and specific gravity at age 80?

Solution

Note: The appendix contains the answer.

2.1.02 Hydrostatic Pressure

Pressure represents the ratio of an applied force and the perpendicular surface area over which it acts. As such, the magnitude of pressure P is equal to the magnitude of the perpendicular force F per area A:

$$P = \frac{F}{A}$$

The SI unit of pressure is the pascal (Pa), which is defined as 1 newton (N) per m²:

$$1 \; \text{Pa} \equiv 1 \; \frac{\text{N}}{\text{m}^2}$$

In simple mechanical systems, pressure is often caused by forceful contact between two solid objects. Consequently, the pressure experienced by any object within a two-object system is directly proportional to the magnitude of the force between the objects and inversely proportional to the area over which the contacting objects meet.

For example, Figure 2.2 shows a machine applying a force of 40 kN on a bone sample.

If the bone sample has a uniform cross-sectional area of 20 mm², the pressure exerted on the bone is calculated as:

$$P = \frac{40 \text{ kN}}{20 \text{ mm}^2} = 2.0 \ \frac{\text{kN}}{\text{mm}^2}$$

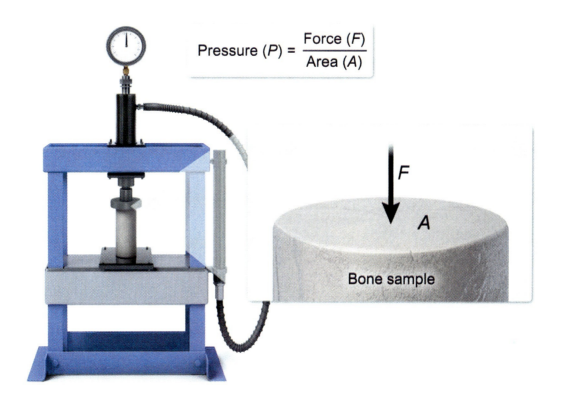

Figure 2.2 Compressing a bone sample.

Applying the unit conversion 1 kN = 1,000 N:

$$P = \left(2.0 \ \frac{\cancel{\text{kN}}}{\text{mm}^2}\right)\left(\frac{1,000 \text{ N}}{1 \ \cancel{\text{kN}}}\right)$$

Furthermore, because 1 m = 1,000 mm:

$$1 \text{ m}^2 = (1,000)^2 \text{ mm}^2 = 1 \times 10^6 \text{ mm}^2$$

Hence:

$$P = \left(2.0 \ \frac{\text{kN}}{\cancel{\text{mm}^2}}\right)\left(\frac{1,000 \text{ N}}{1 \text{ kN}}\right)\left(\frac{1 \times 10^6 \ \cancel{\text{mm}^2}}{1 \text{ m}^2}\right)$$

Therefore, the pressure exerted on the sample of bone is:

$$P = 2.0 \times 10^9 \ \frac{\text{N}}{\text{m}^2} = 2.0 \times 10^9 \text{ Pa}$$

Hydrostatic pressure is the pressure exerted by the weight of a fluid. As shown in Figure 2.3, the hydrostatic pressure P exerted at a point in a fluid is the product of the fluid's density ρ, the gravitational acceleration g, and the height h of the fluid above the point:

$$P = \rho g h$$

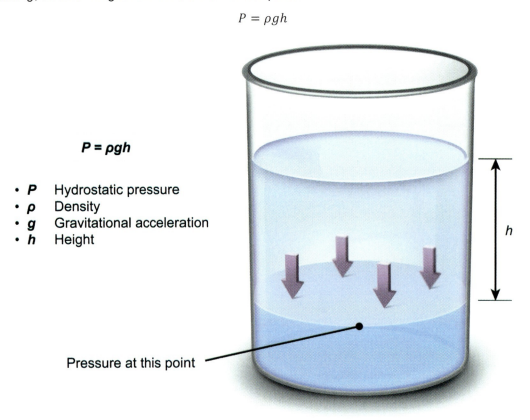

Figure 2.3 Hydrostatic pressure in a fluid.

For example, the pressure experienced by a diver increases with depth because the amount of water directly above them increases (Figure 2.4).

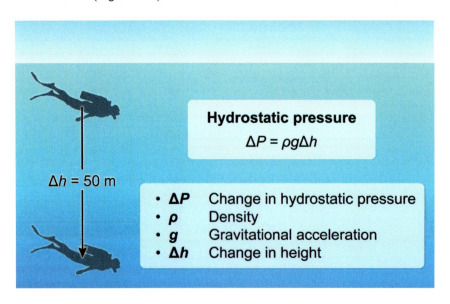

Figure 2.4 The change in hydrostatic pressure on a diver.

During the diver's descent, h increases by 50 m. Given the density of water (ρ = 1,000 kg/m³), the value for the gravitational acceleration (g = 10 m/s²), and 50 m for Δh, the change in hydrostatic pressure ΔP is:

$$\Delta P = \rho g \Delta h$$

$$\Delta P = \left(1{,}000 \ \frac{\text{kg}}{\text{m}^3}\right)\left(10 \ \frac{\text{m}}{\text{s}^2}\right)(50 \text{ m})$$

$$\Delta P = 500{,}000 \ \frac{\text{N}}{\text{m}^2} = 500{,}000 \text{ Pa}$$

Therefore, the change in hydrostatic pressure on the diver expressed in scientific notation is 5×10^5 Pa.

Atmospheric pressure P_{atm} refers to the pressure that gas molecules in the atmosphere exert on all terrestrial objects (eg, bodies of water, living beings). At sea level, P_{atm} is approximately equal to 101,000 Pa. Other approximate values for P_{atm} in equivalent units include:

$$P_{\text{atm}} = 1 \text{ atm} \approx 101{,}000 \text{ Pa} \approx 760 \text{ mmHg}$$

Hence, the absolute pressure P at a depth h in a fluid that is open to the atmosphere is the sum of P_{atm} and the pressure due to the fluid:

$$P = P_{\text{atm}} + \rho g h$$

A **simple fluid barometer**, like the one shown in Figure 2.5, is a device that can be used to measure P_{atm}.

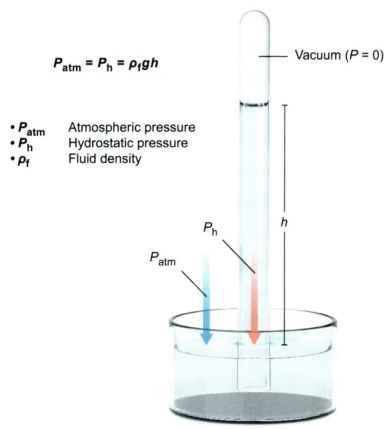

Figure 2.5 A simple barometer is a device used to measure atmospheric pressure.

The hydrostatic pressure P_h generated by the fluid column inside the barometer is a reliable estimate of local P_{atm}:

$$P_{\text{atm}} = P_h$$

The value of P_h itself is related to the fluid density ρ_f, the gravitational acceleration g, and the height h of the fluid column. Therefore, P_{atm} can be quantified in terms of the variables influencing P_h:

$$P_{atm} = P_h = \rho_f g h$$

For example, a barometer filled with mercury (ρ_{Hg} = 13,500 kg/m³) at a standard pressure of 1 atm produces a column of height h:

$$h = \frac{P_{atm}}{\rho_f g} \approx \frac{(1 \text{ atm})}{\left(13{,}500 \; \frac{\text{kg}}{\text{m}^3}\right)\left(10 \; \frac{\text{m}}{\text{s}^2}\right)} = \frac{(101{,}000 \text{ Pa})}{\left(13{,}500 \; \frac{\text{kg}}{\text{m}^3}\right)\left(10 \; \frac{\text{m}}{\text{s}^2}\right)} = 0.75 \text{ m}$$

Hence, an atmospheric pressure of 1 atm is approximately equivalent to the pressure at the bottom of a column of mercury with a height of 750 mm.

For an enclosed fluid, an **external pressure** applied to the fluid is uniformly **transmitted to all surfaces**. This principle is known as **Pascal's law**.

For example, the pressure applied downward on the piston in the left cylinder in Figure 2.6 is transmitted by the fluid to all surfaces, including the piston in the right cylinder.

Equating the pressure P_1 applied by the left cylinder and the pressure P_2 experienced by the right cylinder yields:

$$P_1 = P_2$$

$$\frac{F_1}{A_1} = \frac{F_2}{A_2}$$

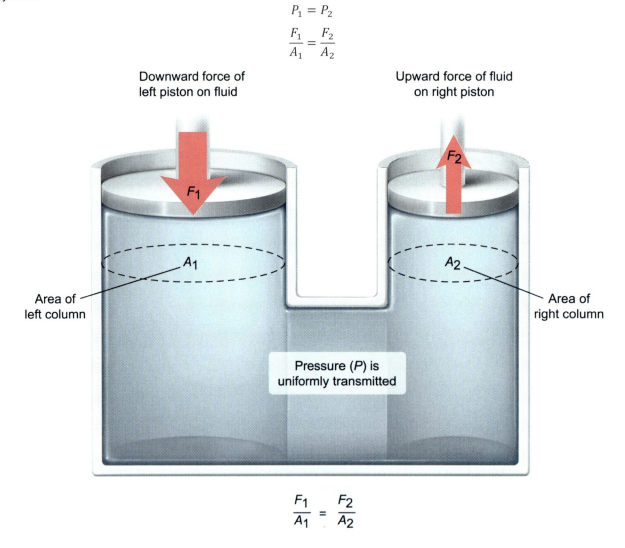

Figure 2.6 The pressure of an enclosed fluid is transmitted uniformly to all surfaces.

For two equal pressures, a greater force is exerted over a proportionally larger area (and vice versa). Rearranging the above equation as a ratio of the force F_2 exerted by the left piston to the force F_1 exerted on the right piston:

$$\frac{F_2}{F_1} = \frac{A_2}{A_1}$$

The ratio of the forces is equal to the ratio of their respective areas. If A_2 is half the area of A_1 (ie, $A_2 = \frac{A_1}{2}$), then $F_2 = \frac{F_1}{2}$, corresponding to a 1:2 ratio between the force on the right piston and the force on the left piston.

 Concept Check 2.3

The systolic blood pressure reading for a patient corresponds to a fluid height of 120 mm of mercury in a simple barometer. If the density of mercury is 13,500 kg/m³, what is the pressure at the surface of the mercury in the barometer?

Solution

Note: The appendix contains the answer.

 Concept Check 2.4

Two syringes filled with water are connected by a tube. The first syringe has a plunger area A_1 of 1 cm². A downward force of 5 N on the first plunger produces an upward force of 20 N on the second plunger. What is the plunger area A_2 of the second syringe?

Solution

Note: The appendix contains the answer.

2.1.03 Buoyancy and Archimedes' Principle

The previous section showed that the pressure created by a fluid at rest is linearly dependent on the depth of the fluid h. This **hydrostatic pressure** is exerted perpendicularly to all surfaces, resulting in horizontal forces applied to a submerged object's sides and vertical forces applied to its top and bottom. Hence, as shown in Figure 2.7, the upward pressure on the bottom of the object (which is at a greater depth) is greater than the downward pressure exerted on the top of the object (which is closer to the fluid's surface).

The difference ΔP between the pressures exerted at the top and bottom of an object is equal to the product of the fluid's density ρ, the gravitational acceleration g, and the difference in depths:

$$\Delta P = \rho g (h_2 - h_1)$$

$P = \rho g h$
$F = \rho g h\, A$

- P Pressure
- F Force on box sides
- ρ Fluid density
- g Gravitational acceleration
- h Depth of fluid
- A Area of box sides

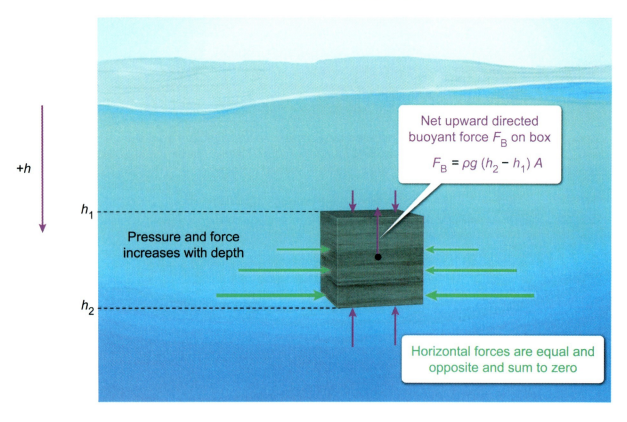

Figure 2.7 Pressure exerted on a box submerged in a fluid.

Pressures directed horizontally on the sides of the object are equal and opposite, and therefore sum to zero. Consequently, a net upward **buoyant force** F_B exists on the object proportional to ΔP. Multiplying ΔP by the area of the top and bottom surfaces gives:

$$F_B = \Delta P \cdot A$$

$$F_B = \rho g (h_2 - h_1) A$$

The product of the area of the top and bottom surfaces and the difference $h_2 - h_1$ is equal to the volume of the box. In general, **Archimedes' principle** states that F_B is equal to the product of ρ, the volume V of the fluid displaced by the object, and g (see Figure 2.8):

$$F_B = \rho V g$$

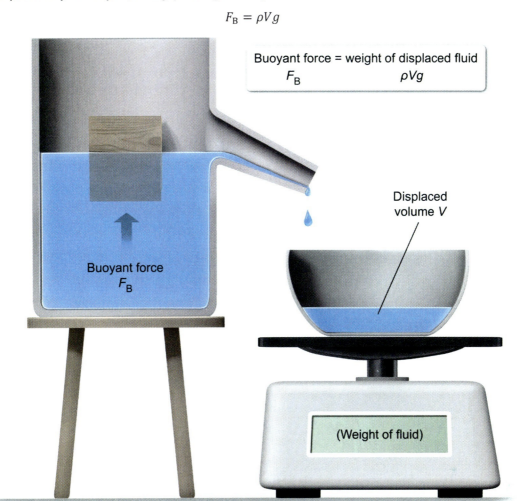

ρ = density; g = gravitational acceleration.

Figure 2.8 Archimedes' principle.

Note that when the object is fully submerged, the buoyant force is always *constant* and does not depend on the object's depth in the fluid.

In addition to the buoyant force exerted on the object, the object's weight F_g always acts downward with a magnitude equal to the product of the object's density ρ_{object}, g, and the object's volume V_{object}:

$$F_g = \rho_{object} \cdot V_{object} \cdot g$$

When an object is fully submerged in the fluid, $V = V_{object}$ and the object's behavior is determined by the relationship between ρ_{object} and fluid density ρ. Three cases of ρ_{object} are shown in Figure 2.9.

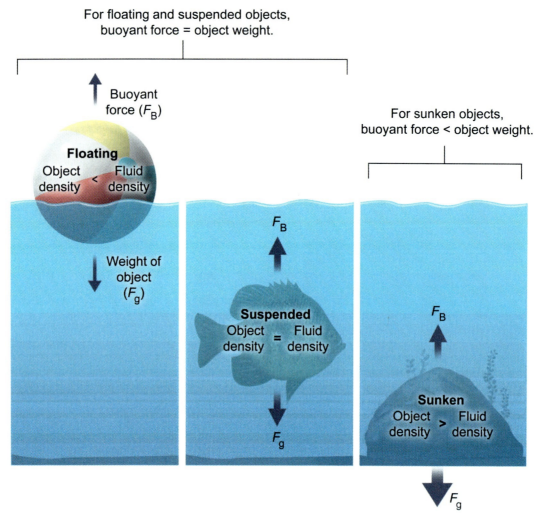

Figure 2.9 An object floats, is suspended, or sinks depending on how its density compares to the fluid density.

In the first case, when $\rho_{object} < \rho$, F_g is then less than F_B and the object rises in the fluid until it **floats**, with some portion of the object protruding above the surface:

$$\rho_{object} < \rho \qquad \Rightarrow \qquad F_g < F_B$$

In the second case, when $\rho_{object} = \rho$, F_g then equals F_B and the object is **suspended** in the fluid:

$$\rho_{object} = \rho \qquad \Rightarrow \qquad F_g = F_B$$

Finally, when $\rho_{object} > \rho$, F_g is greater than F_B and the object immediately **sinks** in the fluid:

$$\rho_{object} > \rho \qquad \Rightarrow \qquad F_g > F_B$$

 Concept Check 2.5

A 100 kg crate is dropped into a river and sinks to the bottom. However, a 10,000 kg boat floats on the surface. How is the boat able to float?

Solution

Note: The appendix contains the answer.

One application of Archimedes' principle is that the **apparent weight** of an object in air differs from its weight when immersed in a fluid. For example, when an object is placed on a submerged scale (Figure 2.10), three forces act on the object: the upward tension T, the upward force F_B, and the downward force F_g. In equilibrium:

$$T + F_B - F_g = 0$$

The apparent weight measured by the scale is T, hence:

$$T = F_g - F_B = \text{apparent weight}$$

For example, a person with a weight of 800 N in air has an apparent weight of 40 N when submerged in a fluid due to the 760 N upward buoyant force (see Figure 2.10).

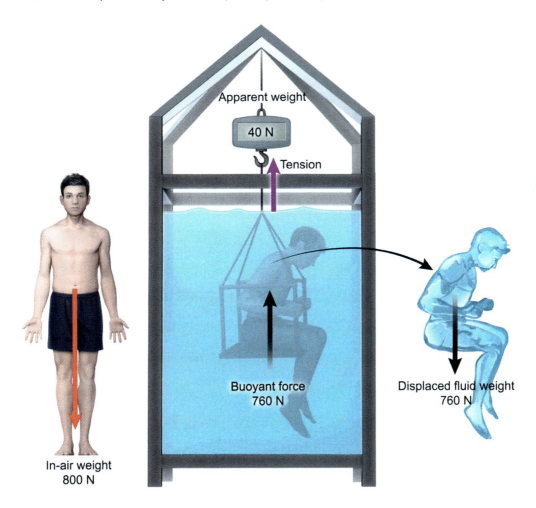

Figure 2.10 Apparent weight of a person immersed in a fluid.

When an object floats on the surface of a fluid, it is in static equilibrium, with $F_g = F_B$, as shown in Figure 2.11.

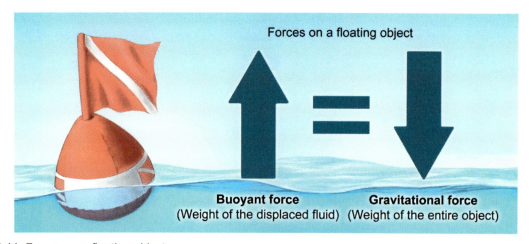

Figure 2.11 Forces on a floating object.

The fraction of the object that is submerged can be determined from the equilibrium condition. In terms of the volume of the fluid displaced V_{fluid} and the total object volume V_{object}:

$$F_B = F_g$$

$$\rho_{fluid} \cdot V_{fluid} \cdot g = \rho_{object} \cdot V_{object} \cdot g$$

$$\rho_{fluid} \cdot V_{fluid} = \rho_{object} \cdot V_{object}$$

Rearranging the above equation yields a relationship between the ratios of the volumes and the ratio of densities, which is the fraction submerged (see Figure 2.12):

$$\frac{\rho_{object}}{\rho_{fluid}} = \frac{V_{fluid}}{V_{object}}$$

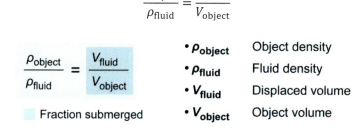

- ρ_{object} Object density
- ρ_{fluid} Fluid density
- V_{fluid} Displaced volume
- V_{object} Object volume

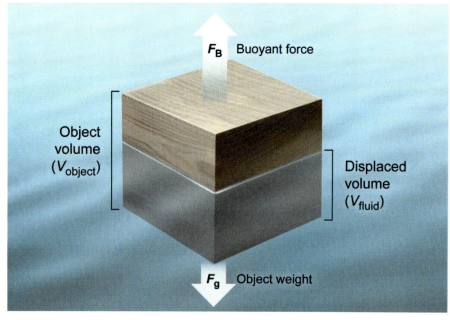

Figure 2.12 Relationship between density and volume for a floating object.

Chapter 2: Fluid Dynamics

> ✓ **Concept Check 2.6**
>
> An iceberg has about 10% of its volume above the surface of the ocean. What is the density of the ice?
>
> **Solution**
>
> *Note: The appendix contains the answer.*

2.1.04 Surface Tension and Capillary Action

For liquids, attractive forces also arise due to the forces between the molecules within the liquid itself, or between the molecules in the liquid and those of its surroundings. These **intermolecular forces** are categorized as either cohesive or adhesive.

Cohesive forces are attractive forces that exist between molecules of the same type and are often responsible for holding a liquid together. Similarly, **adhesive forces** are attractive forces between molecules of different types. For example, water droplets cling to the walls of a bathroom shower due to the adhesive force between the water molecules and the molecules in the shower tile (Figure 2.13).

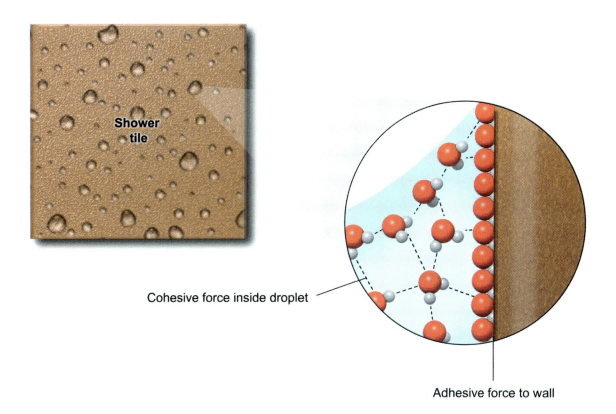

Figure 2.13 Cohesive and adhesive forces for a water droplet.

Inside a liquid, the net cohesive force is zero because each molecule is surrounded by other molecules of the same substance. However, near the surface of the fluid the cohesive force pulls molecules inward, creating **surface tension**. Surface tension acts to decrease the surface area of a liquid to its minimum possible value.

This behavior has two primary effects on the liquid. First, for a fixed volume of liquid, greater surface tension is associated with a smaller surface area of a droplet (ie, the droplet becomes more spherical, Figure 2.14).

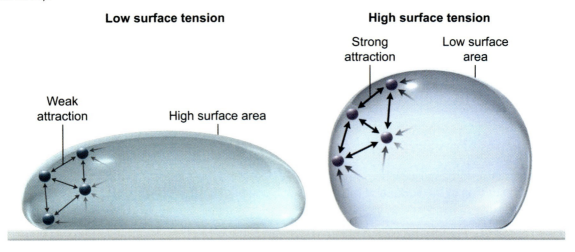

Figure 2.14 For a liquid droplet of fixed volume, greater surface tension is associated with reduced surface area.

Second, the liquid surface behaves like an elastic membrane, with a restoring force analogous to the spring force discussed in Concept 1.3.06. When an object (eg, an insect, a needle) is placed on the surface, it distorts the surface from its equilibrium shape, but the restoring forces trying to minimize the surface area are able to support the weight of the object (Figure 2.15).

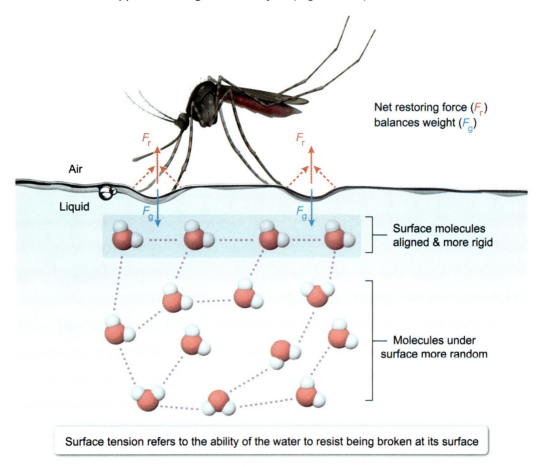

Figure 2.15 Surface tension and the restoring forces due to an insect at rest on the surface of water.

The effects of surface tension are also important to the process of breathing. Due to the hydrogen bonds between the water molecules lining the alveolar sacs, surface tension is significant on alveolar surfaces in the lungs. This restoring force exerts a collapsing pressure on the alveoli, allowing one to exhale without use of the muscles.

Inhalation, which involves filling the alveoli with air and increasing their surface area, is achieved with the help of a chemical called *surfactant*, which decreases the alveoli surface tension. Many breathing problems occur due to the disruption of this surfactant, which makes it difficult to inflate the lungs.

Finally, when a narrow tube is inserted into a liquid, a phenomenon known as **capillary action** occurs. The competition between the surface tension of the liquid and the adhesive forces between the liquid and the tube walls either propels the liquid upward in the tube or suppresses it.

When adhesion overcomes surface tension (eg, water), the liquid rises in the tube and its meniscus curves upward (concave). However, when surface tension is greater than adhesion (eg, mercury) the capillary action suppresses the liquid in the tube and its meniscus curves downward (convex) (Figure 2.16).

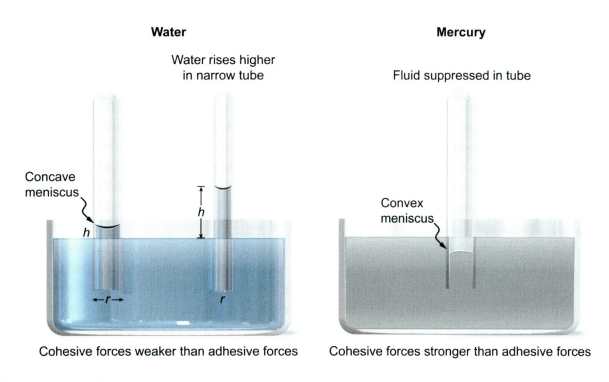

Figure 2.16 Capillary action for water and mercury.

The height *h* of the liquid in a tube due to capillary action is inversely proportional to the tube's radius *r*:

$$h \propto \frac{1}{r}$$

Concept Check 2.7

A student notices that after detergent is added to water, a needle can no longer be supported by the surface of the water. Why does the needle fall into the water?

Solution

Note: The appendix contains the answer.

Lesson 2.2
Fluids in motion

Introduction

This lesson discusses the properties of fluids in motion. The way fluids flow can be classified as either ideal or nonideal. The flow of an ideal fluid through a conduit is very different from the flow of a nonideal fluid. This lesson describes how the continuity equation and Bernoulli's equation for ideal flows can be used to solve problems on the exam. Finally, the physics of nonideal flows is then applied for real-world examples, such as blood flow in the body.

2.2.01 Ideal vs Nonideal Flow

In the study of fluid dynamics, it is important to understand the concepts of **ideal and nonideal fluids**. An ideal fluid exhibits three unique properties. First, an ideal fluid has zero **viscosity**, which refers to a fluid's innate tendency to resist flow (see Figure 2.17). The internal friction that exists between the molecules of a nonideal fluid is not present (ie, the fluid is ideal).

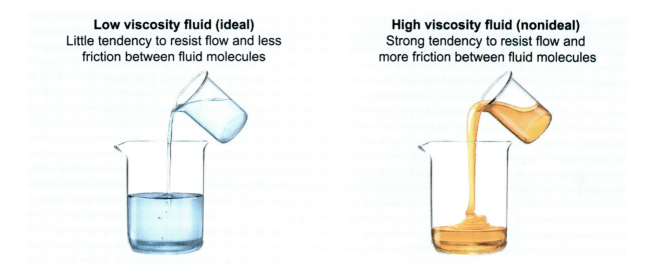

Figure 2.17 Ideal fluids are characterized by their low viscosity whereas nonideal fluids have high viscosity.

Second, an ideal fluid is **incompressible**, which means that unlike a nonideal fluid, it has a constant density that is independent of the pressure applied (Figure 2.18). Water and air are both examples of fluids that can approximate ideal behavior under generalized conditions.

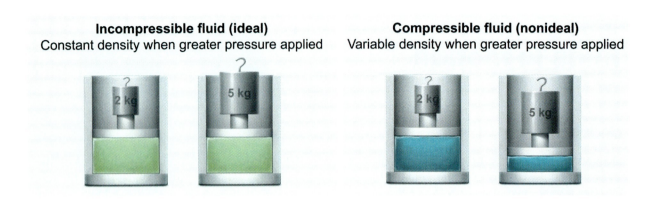

Figure 2.18 Ideal fluids are incompressible whereas nonideal fluids exhibit a variable density and can be compressed.

Third, an ideal fluid always flows smoothly and easily (ie, **laminar flow**). This means that the ideal fluid moves in a straight line and without any loss of energy as shown in Figure 2.19. However, a nonideal fluid can experience **turbulent flow** where the fluid flow becomes chaotic, changing directions and forming swirls.

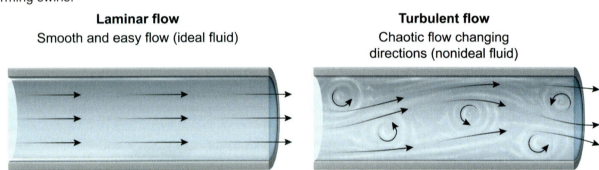

Figure 2.19 Ideal fluids experience laminar flow whereas nonideal fluids may experience turbulent flow.

The behavior of an ideal fluid serves as a model for the application of **Bernoulli's equation**, which mathematically describes the conservation of energy in fluids (Concept 2.2.03).

2.2.02 Continuity Equation

When a fluid flows through a conduit such as a pipe or a blood vessel, the quantity of mass entering and exiting the system must remain constant. When the fluid is ideal, conservation of mass implies that the volumetric flow rate Q, which is equal to the product of the conduit's **cross-sectional area** A and the **fluid velocity** v, must remain constant:

$$Q = Av = \text{constant}$$

Consequently, the flow rate, Q_1 and Q_2, at any two points along the conduit's path can be equated, yielding the **continuity equation** (Figure 2.20):

$$Q_1 = Q_2$$
$$A_1 v_1 = A_2 v_2$$

When the cross-sectional area of the conduit varies, the speed of the fluid must vary inversely to keep the total volume of fluid flow constant.

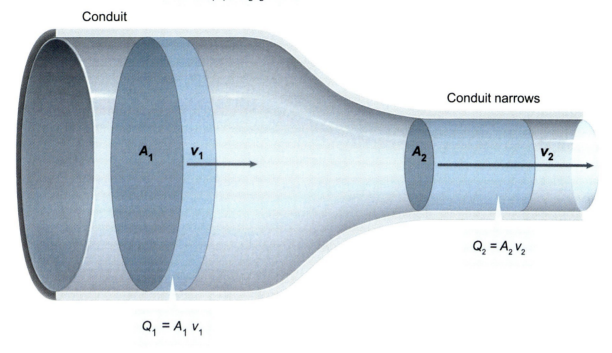

Figure 2.20 When a conduit narrows, the speed of the fluid must increase to keep the volumetric flow rate constant.

It is important to note that this form of the continuity equation is only applicable to ideal fluids. In cases of nonideal fluid flow, Q is not necessarily constant and another equation may be needed to accurately describe the flow of the fluid.

 Concept Check 2.8

The diameter of a segment of an artery is reduced by a factor of four due to an obstruction. Assume blood is an ideal fluid and that flow through the artery follows the continuity equation. Compared to an unobstructed segment of the artery, how does the velocity of blood in the obstructed segment change?

Solution

Note: The appendix contains the answer.

2.2.03 Bernoulli's Equation

To further describe the flow of an ideal fluid between two points A and B in a conduit (such as a pipe or blood vessel), **Bernoulli's equation** is often used. In a closed system, the total fluid energy remains constant between points A and B.

The total energy E of the fluid has three components:

- The potential energy of the fluid (which is related to its height h).
- The kinetic energy of the fluid (which is related to the fluid velocity v).
- The pressure P of the fluid.

The **potential energy** of a given volume of fluid is equal to the product of the **fluid density** ρ, the **gravitational acceleration** g (ie, the free-fall acceleration), and its height h above a reference level:

$$\text{Potential energy per unit volume} = \rho g h$$

Moreover, the **kinetic energy** of the volume of fluid is equal to half the product of ρ and the square of v:

$$\text{Kinetic energy per unit volume} = \frac{1}{2}\rho v^2$$

Finally, the **pressure** is associated with the work done on each element of the fluid per unit volume:

$$\text{Work per unit volume} = P$$

Setting the sum of these three energy terms at points A and B equal yields Bernoulli's equation (see Figure 2.21).

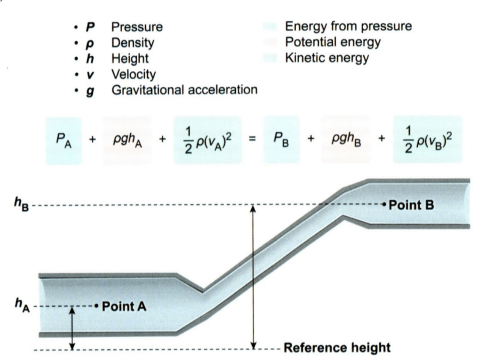

Figure 2.21 Bernoulli's equation applied to a pipe at different points A and B.

When the height of the fluid does not change, the potential energy terms cancel. This results in a special condition of Bernoulli's equation known as *Bernoulli's principle*. The resulting equation describes the **Venturi effect**, in which a decreased pressure is associated with an increased fluid velocity as shown in Figure 2.22.

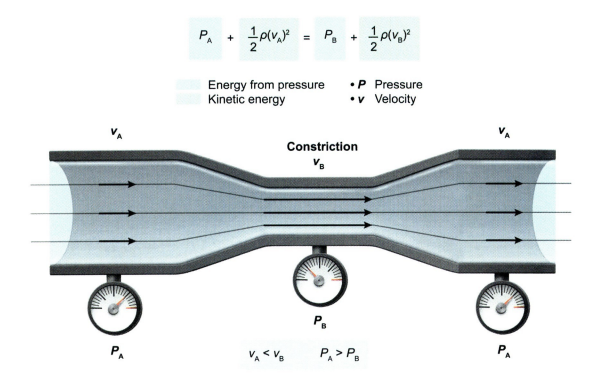

Figure 2.22 The Venturi effect describes the inverse relationship between fluid velocity and pressure when the height remains constant.

 Concept Check 2.9

In the vertical pipe shown below, a liquid flows a distance of 5 cm between two points A and B. The pressure at point A is 6 kPa. What is the pressure when the liquid reaches point B? Assume the liquid's velocity is constant, the density of the liquid is 1,200 kg/m³, and the gravitational acceleration is 10 m/s².

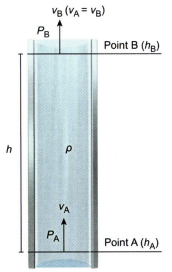

Solution
Note: The appendix contains the answer.

2.2.04 Poiseuille Viscous Flow and Turbulence

One important characteristic of an **ideal fluid** is that it has zero viscosity and flows freely through a conduit, such as a pipe or blood vessel. For example, the water flowing through a garden hose can be treated as an ideal fluid because it experiences very little **resistance** to flow (ie, ideal flow).

However, many real-life fluids are viscous due to internal **friction** forces between the molecules, and they behave like nonideal fluids. These internal forces are **nonconservative**, causing resistance to fluid flow through the conduit.

This resistance to a fluid's flow depends on its viscosity η. The SI unit for η is the product of pascals and seconds, Pa·s. The η for some common fluids are shown in Table 2.1.

Table 2.1 Viscosity of common fluids.

Fluid	Viscosity (η)
Water	0.8 mPa·s
Milk	2 mPa·s
Blood	3 mPa·s
Motor oil	500 mPa·s

Recall from Concept 2.2.02 that in the **continuity equation**, the volumetric flow rate Q of an ideal fluid is equal to the product of the pipe area A and the fluid velocity v:

$$Q = Av = \pi r^2 v$$

However, the flow of a viscous fluid through a pipe follows **Poiseuille's law**, which includes terms for the radius r of the pipe, the change in pressure ΔP along the pipe, the fluid viscosity η, and the length L of the pipe:

$$Q = \frac{\pi r^4 \cdot \Delta P}{8 \eta L}$$

Poiseuille's law only applies to a nonideal fluid under the condition of laminar flow. Laminar flow consists of parallel regions of fluid all flowing in the same direction. However, the velocity of the fluid is not constant because the velocity is greatest at the center of the pipe and slower at the interior surface of the pipe.

For example, suppose that blood flows through a 5 cm segment of a blood vessel with a radius of 0.2 cm and a pressure change of 2 Pa, as shown in Figure 2.23.

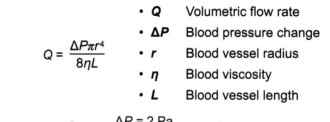

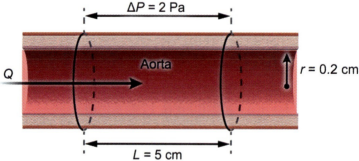

Figure 2.23 Volumetric blood flow through a blood vessel.

The blood flow through a blood vessel follows Poiseuille's law because blood is a viscous fluid:

$$Q = \frac{\pi r^4 \cdot \Delta P}{8 \eta L}$$

From Table 2.1, the viscosity η of blood is 3 mPa·s. The change in blood pressure ΔP = 2 Pa, the vessel length L = 5 cm, and the radius r = 0.2 cm. Substituting these values into the equation for Poiseuille's law yields:

$$Q = \frac{\pi (0.2 \text{ cm})^4 (2 \text{ Pa})}{8(3 \text{ mPa} \cdot \text{s})(5 \text{ cm})}$$

$$Q = \frac{(3.14)(2 \times 10^{-3} \text{ m})^4 (2 \text{ Pa})}{8(3 \times 10^{-3} \text{ Pa} \cdot \text{s})(5 \times 10^{-2} \text{ m})}$$

$$Q = \frac{(3.14)(16 \times 10^{-12})(2)}{8(15 \times 10^{-5})} \approx \frac{100 \times 10^{-12}}{120 \times 10^{-5}}$$

$$Q \approx 0.8 \times 10^{-7} \approx 8 \times 10^{-8} \, \frac{\text{m}^3}{\text{s}}$$

Under some circumstances, the flow of a nonideal fluid can change from laminar flow to turbulent flow. During turbulent flow, the fluid no longer flows in one constant direction down the pipe, but instead some of the fluid may flow backward or swirl within the pipe. Turbulent flow can occur due to **obstructions** within the pipe or when the **velocity** of the fluid is so great that it overcomes the smoothing effect of viscosity. These effects in blood flow are described in Concept 2.2.05.

 Concept Check 2.10

What is the flow rate of water through a pipe with a length of 1 m and a radius of 1 cm when the change in pressure is 10 Pa?

Solution

Note: The appendix contains the answer.

2.2.05 Applications to Blood Circulation

The **cardiovascular system** is a closed loop in which **blood** flows from the heart through a network of blood vessels consisting of the arteries, arterioles, capillaries, venules, and veins. As blood flows through the different vascular regions, the blood pressure varies, as shown in Figure 2.24.

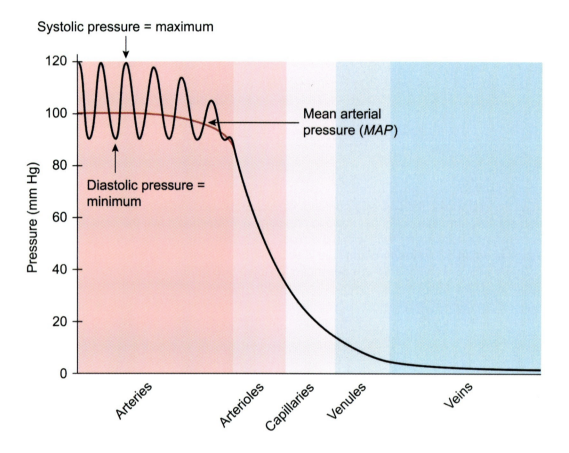

Figure 2.24 Blood pressure in different vascular regions.

In the **arteries**, the blood pressure varies over time due to the periodic contractions of the heart. The maximum blood pressure in the arteries is called the **systolic pressure** P_S, and the minimum blood pressure is called the **diastolic pressure** P_D. The pulse pressure PP is defined as the difference of P_S and P_D:

$$PP = P_S - P_D$$

Another common description of blood pressure is mean arterial pressure MAP, which is determined from P_S and P_D according to the equation:

$$MAP = P_D + \frac{P_S - P_D}{3}$$

Because the cardiovascular system is a closed loop, the volumetric flow rate through each of the vascular regions must be constant. In other words, if 5 L of blood flows through the arteries in 1 min, then 5 L of blood must also flow through the capillaries in 1 min.

Recall from Concept 2.2.02 that the volumetric flow rate Q of an ideal fluid is equal to the product of the pipe's cross-sectional area A and the fluid velocity v:

$$Q = Av = \pi r^2 v$$

The cardiovascular system consists of vessels with different cross-sectional areas, but each vascular region also has a different number *n* of parallel vessels. Thus, the continuity equation calculates blood flow through a vascular region as the product of *n*, *A*, and *v*:

$$Q = nAv = n\pi r^2 v$$

The blood flow through the vasculature depends on the amount of blood pumped by the heart, known as the cardiac output *CO* (Figure 2.25). The *CO* is equal to the product of the **stroke volume** *SV* of the heart and the **heart rate** *HR*:

$$CO = SV \cdot HR$$

The *SV* of the heart is the volume of blood pumped out of the left ventricle during a single heartbeat.

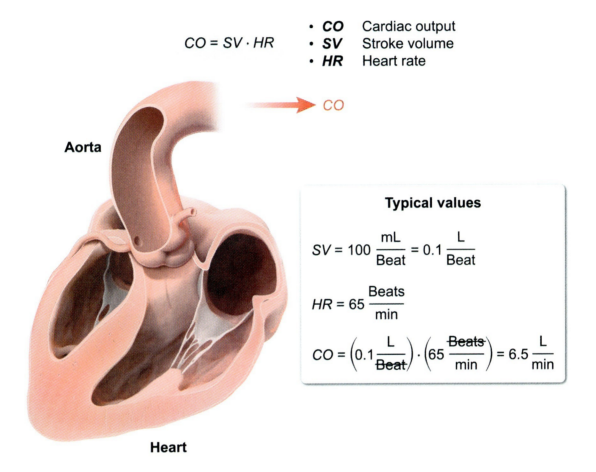

Figure 2.25 Cardiac output depends on the stroke volume and heart rate.

Figure 2.26 is a plot of the volume and pressure in the left ventricle over a single heartbeat, known as a pressure-volume loop, or **PV loop**. The *SV* is equal to the difference between the maximum and minimum volumes of the left ventricle.

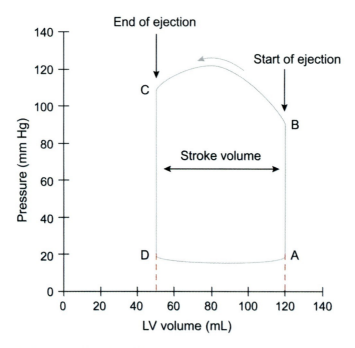

Figure 2.26 PV loop from the left ventricle (LV) of the heart.

The area A_{PV} enclosed by the PV loop represents the product of pressure P and volume V:

$$A_{PV} = PV$$

Recall from Concept 2.1.02 that P represents the ratio of a force F and the area A over which it is applied:

$$P = \frac{F}{A}$$

Substituting this relationship into the equation for A_{PV} yields:

$$A_{PV} = \left(\frac{F}{A}\right)V = F\frac{V}{A}$$

Because the units for V and A are m³ and m², respectively, the ratio of V and A has the units of meters, which corresponds to distance d:

$$\frac{V}{A}\left[\frac{m^3}{m^2}\right] = d\,[m]$$

Therefore, A_{PV} is equal to the product of F and d:

$$A_{PV} = F\frac{V}{A} = Fd$$

Also recall from Concept 1.5.01 that work W is equal to the product of F and d. Hence, A_{PV} represents the work done by the heart during a single heartbeat:

$$A_{PV} = Fd = W$$

Blood flow through the vasculature can only be calculated with the continuity equation if blood is assumed to be an **ideal fluid**. However, blood is a **viscous** fluid and follows **Poiseuille's law**, as shown in Figure 2.23. Blood flow is typically **laminar**, but flow can become turbulent when it exceeds a **critical velocity** v_c, which depends on the fluid viscosity η, the **Reynolds number** Re, fluid density ρ, and vessel diameter D:

$$v_c = \frac{\eta \cdot Re}{\rho D}$$

The Reynolds number is a unitless value that depends on the physical properties of the vessel walls and any obstructions within the vessel itself.

For example, blood has a density of 1,000 kg/m³, viscosity of 0.003 kg/(m·s), and becomes turbulent at a Reynolds number of 5,000. The critical velocity of blood flowing through a vessel with a diameter of 2 cm is calculated from the equation above, yielding:

$$v_c = \frac{\left(0.003 \frac{kg}{m \cdot s}\right)(5{,}000)}{\left(1{,}000 \frac{kg}{m^3}\right)(2 \text{ cm})}$$

Converting the units for D into meters and solving gives:

$$v_c = \frac{15 \frac{kg}{m \cdot s}}{\left(1{,}000 \frac{kg}{m^3}\right)(2 \times 10^{-2} \text{ m})} = \frac{15 \frac{1}{m \cdot s}}{20 \frac{1}{m^3}} = 0.75 \frac{m}{s}$$

Therefore, the blood flow is laminar up to a maximum velocity of 0.75 m/s. If the blood velocity exceeds 0.75 m/s, the blood flow becomes turbulent.

Concept Check 2.11

A single arteriole with a cross-sectional area of 2×10^{-9} m² feeds 500 capillaries with an average cross-sectional area of 0.1×10^{-9} m². What is the blood velocity in the capillaries if the blood velocity in the arteriole is 1×10^{-1} m/s?

Solution

Note: The appendix contains the answer.

Chapter 2: Fluid Dynamics

END-OF-UNIT MCAT PRACTICE

Congratulations on completing **Unit 2: Fluids**.

Now you are ready to dive into MCAT-level practice tests. At UWorld, we believe students will be fully prepared to ace the MCAT when they practice with high-quality questions in a realistic testing environment.

The UWorld Qbank will test you on questions that are fully representative of the AAMC MCAT syllabus. In addition, our MCAT-like questions are accompanied by in-depth explanations with exceptional visual aids that will help you better retain difficult MCAT concepts.

TO START YOUR MCAT PRACTICE, PROCEED AS FOLLOWS:

1) Sign up to purchase the UWorld MCAT Qbank
   IMPORTANT: You already have access if you purchased a bundled subscription.
2) Log in to your UWorld MCAT account
3) Access the MCAT Qbank section
4) Select this unit in the Qbank
5) Create a custom practice test

Unit 3 Electrostatics and Circuits

Chapter 3 Electricity and Magnetism

3.1 Electric Charge and Force

    3.1.01   Concept of Electric Charge
    3.1.02   Electrostatic Force and Coulomb's Law
    3.1.03   Electric Field
    3.1.04   Electric Potential Energy and Voltage

3.2 Flowing Charge

    3.2.01   Definition of Current
    3.2.02   Conductivity and Resistivity
    3.2.03   Circuit Basics: Resistance and Ohm's Law
    3.2.04   Series and Parallel Circuits
    3.2.05   Meters

3.3 Capacitance

    3.3.01   Definition of Capacitance
    3.3.02   Energy of a Charged Capacitor
    3.3.03   Parallel Plate Capacitors
    3.3.04   Capacitors in Series and Parallel

3.4 Magnetism

    3.4.01   Magnets and the Magnetic Field
    3.4.02   Lorentz Force

Lesson 3.1

Electric Charge and Force

Introduction

All matter contains electric charges because every atom contains both positive and negative charges in the form of protons and electrons. The attractive force between protons and electrons holds atoms together to form molecules. Molecules interact with one another electrically to produce solids and liquids like bone and blood.

This lesson explores the fundamental properties of charge and electrostatic force. Among these properties are the facts that electric charge comes in discrete quantities and is conserved. Next, the lesson investigates the force law governing the interaction between charges and develops the concepts of the electric field, electric potential energy, and voltage.

3.1.01 Concept of Electric Charge

Electric charge is a fundamental property of matter that influences many phenomena. Several conventions exist for quantifying the magnitude of charge associated with an object. Many of these conventions relate to the basic structure of the atom, which consists of positively charged **protons**, uncharged **neutrons**, and negatively charged **electrons**, as seen in Figure 3.1.

The SI unit of electric charge is the coulomb (C), where +1 C is equivalent in charge to approximately 6.24×10^{18} protons, and −1 C is equivalent in charge to approximately 6.24×10^{18} electrons:

$$+1\,C = 6.24 \times 10^{18} \text{ protons} \qquad -1\,C = 6.24 \times 10^{18} \text{ electrons}$$

The charge magnitudes of one electron and one proton are not only equal but are also the smallest unit of charge, known as the **elementary charge** q_e:

$$+1\,q_e = 1 \text{ proton charge} = \frac{+1\,C}{6.24 \times 10^{18}} = +1.60 \times 10^{-19}\,C$$

Similarly, the magnitude of charge on one electron is also the elementary charge:

$$-1\,q_e = 1 \text{ electron charge} = \frac{-1\,C}{6.24 \times 10^{18}} = -1.60 \times 10^{-19}\,C$$

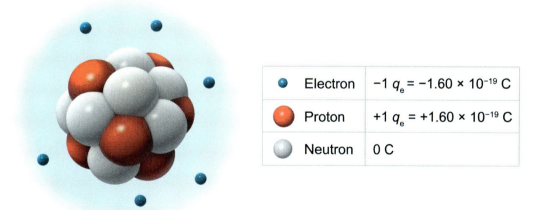

Figure 3.1 Elementary charge in an atom.

Hence, the charge Q on an object is some integer multiple *n* of the elementary charge:

$$Q = nq_e$$

The law of **conservation of charge** states that electric charge is neither created nor destroyed. A neutral object has an *equal* amount of positive and negative charge. Because protons in atoms are confined to the nuclei, ordinary materials typically gain or lose charge via the movement of electrons.

When two objects are rubbed together, **electrons** can transfer from one object to the other. Objects that gain electrons have an excess negative charge (ie, become negatively charged). The number of excess electrons on a negatively charged object is:

$$n = \frac{Q}{q_e}$$

Similarly, objects that lose electrons have an equal amount of excess positive charge (ie, become positively charged). The total charge of the system remains constant, but the charge is redistributed among the objects in the system.

For example, when a neutral plastic rod (shown in Figure 3.2) is rubbed against neutral animal fur, electrons are transferred from the fur to the rod, giving the rod a charge of −1.5 × 10⁻⁶ C and the animal fur a charge of +1.5 × 10⁻⁶ C. The number of electrons transferring in the process is calculated as:

$$n = \frac{Q}{q_e} = \frac{-1.5 \times 10^{-6} \, \cancel{C}}{-1.60 \times 10^{-19} \, \cancel{C}/\text{electron}} \approx 9 \times 10^{12} \text{ electrons}$$

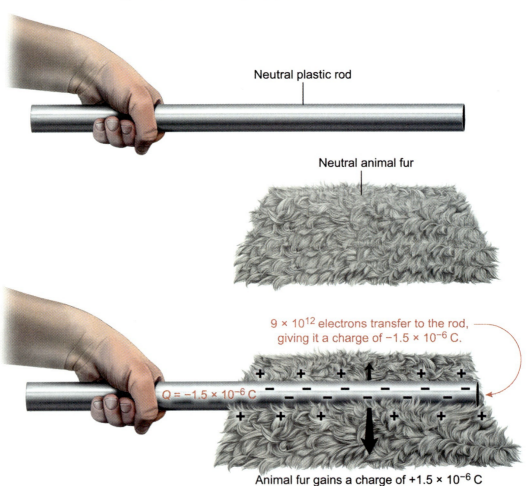

Figure 3.2 A neutral plastic rod becomes negatively charged after being rubbed on animal fur.

Chapter 3: Electricity and Magnetism

Concept Check 3.1

−2Q   **+6Q**

Sphere 1   Sphere 2

There are excess electrons on the surface of conducting sphere 1, giving it a charge of −2Q, as shown above. Conducting sphere 2 with a charge of +6Q is briefly brought into contact with sphere 1 and then separated. The two spheres end up with equal charge. What is the charge on sphere 1 and sphere 2 after contact?

Solution

Note: The appendix contains the answer.

3.1.02 Electrostatic Force and Coulomb's Law

The previous section established that two types of electric charge exist: positive and negative. The sign (positive or negative) and quantity of electric charge influence how charged particles interact. Every electric charge exerts an electrostatic force on any other charge nearby.

Furthermore, the electrostatic force is a vector quantity and acts along a radial direction between the centers of the charges. Electric charges with the *same* sign (ie, like charges) exert *repulsive* forces against each other whereas charges with *opposite* signs (ie, unlike charges) exert *attractive* forces on one another, as shown in Figure 3.3.

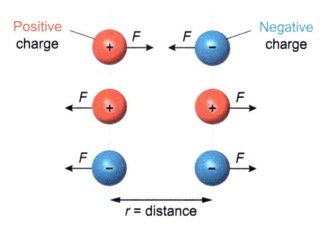

Figure 3.3 Electrostatic forces on like and unlike charges.

Coulomb's law states that the magnitude of the electrostatic force F_E between two charges is directly proportional to the quantity of charge on each particle (q_1, q_2) but inversely proportional to the square of the distance r separating the charges. This relationship can be expressed as Coulomb's law equation (where $k = 8.99 \times 10^9$ N·m²·C⁻²):

$$F_E = k \frac{q_1 q_2}{r^2}$$

For example, Figure 3.4 shows an attractive force between a particle with a charge of +6.0 nC and a particle with a charge of −3.0 nC separated by 50 cm.

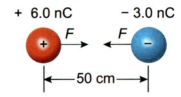

Figure 3.4 Calculating the magnitude of the attractive electrostatic force between two opposite charges.

Substituting the values of q_1, q_2, and r into Coulomb's law equation:

$$F_E = \left(8.99 \times 10^9 \ \frac{\text{N·m}^2}{\text{C}^2}\right) \frac{(+6.0 \ \text{nC})(-3.0 \ \text{nC})}{(50 \ \text{cm})^2}$$

According to the list of metric prefixes, 1 nC = 10⁻⁹ C and 1 cm = 10⁻² m. Hence:

$$F_E = \left(8.99 \times 10^9 \ \frac{\cancel{\text{N·m}^2}}{\cancel{\text{C}^2}}\right) \frac{(+6.0 \times 10^{-9} \ \cancel{\text{C}})(-3.0 \times 10^{-9} \ \cancel{\text{C}})}{(0.5 \ \cancel{\text{m}})^2}$$

$$F_E = -6.5 \times 10^{-7} \ \text{N}$$

The *negative* sign on F_E indicates that the electrostatic force between the particles is *attractive*.

On the other hand, Figure 3.5 shows a repulsive force between a particle two identical particles, each with a charge of +6.0 nC, separated by 50 cm.

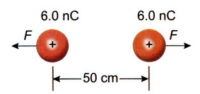

Figure 3.5 Calculating the magnitude of the repulsive electrostatic force between two positive charges.

Using Coulomb's law equation yields:

$$F_E = k \frac{q_1 q_2}{r^2}$$

$$F_E = \left(8.99 \times 10^9 \ \frac{\cancel{\text{N·m}^2}}{\cancel{\text{C}^2}}\right) \frac{(+6.0 \times 10^{-9} \ \cancel{\text{C}})(+6.0 \times 10^{-9} \ \cancel{\text{C}})}{(0.5 \ \cancel{\text{m}})^2}$$

$$F_E = 1.3 \times 10^{-6} \ \text{N}$$

The *positive* sign on F_E indicates that the electrostatic force is *repulsive*. A positive sign also results when calculating the force between two negatively charged particles, indicating that two particles with the same sign will always produce a repulsive force.

> **✓ Concept Check 3.2**
>
>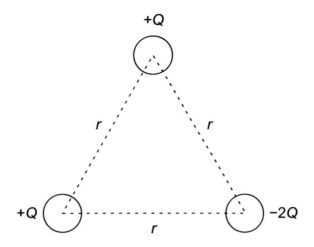
>
> Three charged spheres are each held fixed at a distance *r* from the other two, as shown above. The diameter of each sphere is much smaller than *r*. Compare the magnitude of the electrostatic force exerted by the bottom +Q and −2Q charges on the +Q charge at the top. In what direction is each force?
>
> **Solution**
>
> *Note: The appendix contains the answer.*

3.1.03 Electric Field

As discussed previously, electrically charged particles exert an **electrostatic force** F_E on one another. Every point charge creates a region of influence called an electric field, causing oppositely charged particles to attract each other and like-charged particles to repel each other, as discussed in Concept 3.1.02. As seen in Figure 3.6, the electric field is equal to the ratio of the electrostatic force $\vec{F}_E$ exerted by the field and the charge q in the field:

$$\vec{E} = \frac{\vec{F}_E}{q}$$

The magnitude of an electric field has SI units of **newtons per coulomb** (N/C).

Furthermore, both $\vec{E}$ and $\vec{F}_E$ are vector quantities. When q is *positive*, $\vec{E}$ and $\vec{F}_E$ act in the *same* direction. Alternatively, when q is *negative*, $\vec{E}$ and $\vec{F}_E$ act in *opposite* directions.

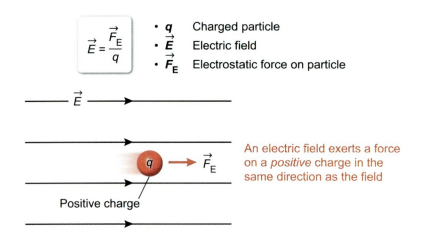

Figure 3.6 A charge q in an electric field $\vec{E}$ is subject to an electrostatic force $\vec{F}_E$.

The equation for the electric field can be rearranged to solve for the electrostatic force $\vec{F}_E$ exerted on a charge q in an electric field:

$$\vec{F}_E = q\vec{E}$$

Electric field lines graphically represent electric fields by indicating the direction and relative strength of $\vec{E}$. Electric field lines always begin at positive charges and end at negative charges, as shown in Figure 3.7.

In other words, electric field lines point in the direction of the force on a positive charge located in the field. Consequently, positive charges accelerate in the same direction as the field lines, and negative charges accelerate in the opposite direction.

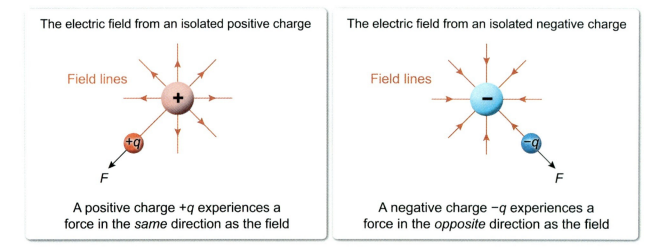

Figure 3.7 Electric field lines surrounding positive and negatively charged particles.

The relative strength of an electric field is represented graphically by the spacing of electric field lines. As seen in Figure 3.8, the field lines are spaced closer together near the charges and farther apart away from the charges. Hence, the electric field is stronger near a single charge and weaker farther away from the charge.

The strength of E can be illustrated by the relative spacing of field lines

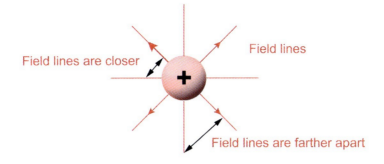

Where field lines are spaced closer together, the field is stronger.

Figure 3.8 The spacing of Electric field lines illustrates the relative strength of E.

☑ **Concept Check 3.3**

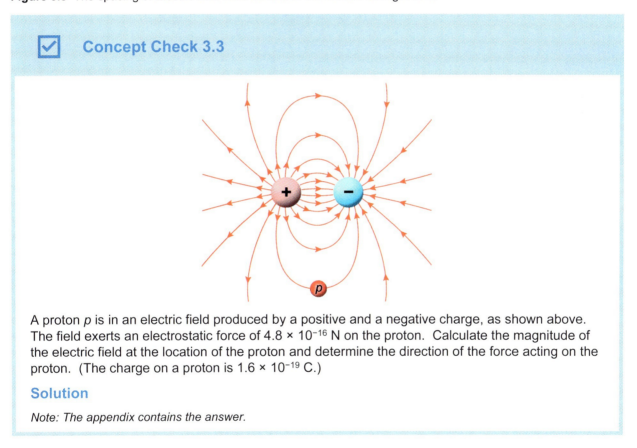

A proton *p* is in an electric field produced by a positive and a negative charge, as shown above. The field exerts an electrostatic force of 4.8 × 10⁻¹⁶ N on the proton. Calculate the magnitude of the electric field at the location of the proton and determine the direction of the force acting on the proton. (The charge on a proton is 1.6 × 10⁻¹⁹ C.)

Solution

Note: The appendix contains the answer.

3.1.04 Electric Potential Energy and Voltage

Concept 3.1.03 referred to the electric field surrounding isolated point charges. In addition, it was shown that the electric field exerts an electrostatic force F_E on charges with magnitude:

$$F_E = qE$$

In Figure 3.9, two oppositely charged conducting parallel plates separated by a distance d produce a uniform electric field in the space between the plates. The top plate is positively charged, meaning that the plate is at a relatively high electric potential. The bottom plate has a negative charge, meaning that it has a relatively low electric potential. Equal spaced downward field lines reveal the uniform electric field that exerts an upward force F_E on a negative charge and a downward force F_E on an equal magnitude positive charge. If free to move, these forces cause the charges to accelerate.

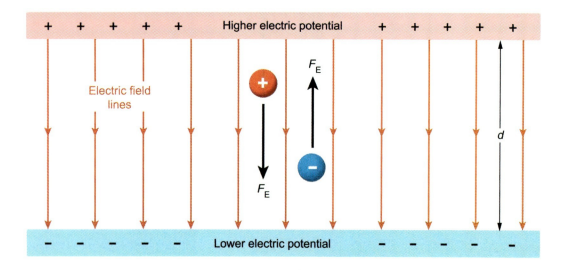

Figure 3.9 Electric field and electric potential for oppositely charged parallel plates.

According to Newton's second law of motion, F_E is equal to the product of the mass m and the constant acceleration a of the charged particle:

$$F_E = ma$$

Rearranging and solving for a yields:

$$a = \frac{F_E}{m}$$

Substituting qE for F_E from the equation for electric field strength results an equation that can be used to calculate the acceleration of a particle with mass m and charge q in and electric field with strength E:

$$a = \frac{qE}{m}$$

For example, an electron with a mass of 9.1 × 10⁻³¹ kg and a charge of −1.6 × 10⁻¹⁹ C in a uniform electric field with a magnitude of 1,000 N/C accelerates with a magnitude calculated as:

$$a = \frac{(1.6 \times 10^{-19} \text{ C})\left(1{,}000 \, \frac{\text{N}}{\text{C}}\right)}{(9.1 \times 10^{-31} \text{ kg})} = 1.8 \times 10^{14} \, \frac{\text{m}}{\text{s}^2}$$

A unit analysis reveals that the units for a are m/s², as expected.

Not only does the electric field shown in Figure 3.9 exert forces on the charges placed in the field causing them to accelerate, but when the charges move, the forces do work W on the charges. The magnitude of **work** done on an electric charge is equal to the product of the electrostatic force F_E and the distance d moved along the electric field:

$$W = F_E \cdot d$$

Substituting the equation above for electrostatic force yields:

$$W = (qE)d$$

A charge moving through an electric field is analogous to a mass moving through a gravitational field, which was described in Concept 1.5.01. Both the electrostatic and gravitational forces are conservative, which implies that work done by the electrostatic force on a charge is *independent of the path* taken by the charge through the electric field.

Hence, the work done depends only on the initial and final positions of the charge and is associated with a change in the **electric potential energy** ΔPE_E:

$$W = \Delta PE_E = qEd$$

The electric potential difference or **voltage** V is defined as the change in electric potential energy per unit charge:

$$V = \frac{\Delta PE_E}{q} = Ed$$

The SI unit for voltage is volts (V), where:

$$1\,\text{V} \equiv 1\,\frac{\text{J}}{\text{C}}$$

As explained in Concept 1.5.04, conservation of energy also applies to the motion of electric charge. The total energy E, which is the sum of the electric potential energy PE_E and kinetic energy KE of the charge, remains constant if the electrostatic force is the only force acting on the charge:

$$E = PE_E + KE = \text{constant}$$

Because E is conserved, the energy of the charge is transformed from PE_E into KE as it accelerates within the electric field:

$$E_\text{initial} = E_\text{final}$$
$$(PE_E + KE)_\text{initial} = (PE_E + KE)_\text{final}$$

where:

$$PE_E = qEd \quad \text{and} \quad KE = \tfrac{1}{2}mv^2$$

 Concept Check 3.4

An electron is released from rest in a uniform electric field with a magnitude of 3,000 N/C and allowed to accelerate. After a distance of 5 cm, how fast is the electron moving? (The mass and charge of an electron are 9.1 × 10⁻³¹ kg and 1.6 × 10⁻¹⁹ C, respectively.)

Solution

Note: The appendix contains the answer.

Lesson 3.2
Flowing Charge

Introduction

This lesson discusses the physics and applications of flowing charge (ie, electricity). After introducing the notion of current and a material's intrinsic resistance to charge flow, the basics of circuits, particularly Ohm's law is addressed. Series and parallel circuit configurations are introduced, and Ohm's law and Kirchhoff's laws are used to develop formulas for how resistors in series and in parallel can be calculated as a total equivalent resistance. The lesson concludes with a brief discussion of the use of ammeters and voltmeters to measure current and voltage in circuit elements.

3.2.01 Definition of Current

The previous lesson described the electric force, electric field, and potential energy associated with static electric charges. However, in most applications, charges move and create a current. A **current** I is defined as the amount of electric charge Q traveling through an object (eg, a wire) per unit time t:

$$I = \frac{Q}{t}$$

The units of current are amperes (A), which is equivalent to coulombs per second (C/s).

Current also has an associated direction. Because positive charges move in the direction of the electric field and from high voltage to low voltage, the convention is that the direction of the current is identical to the direction that positive charges move.

For most states of matter, the only charged particles that travel throughout a material are negatively charged electrons. As a result, the direction of current flow (by convention) and the actual direction of electron flow are typically opposite one another (Figure 3.10).

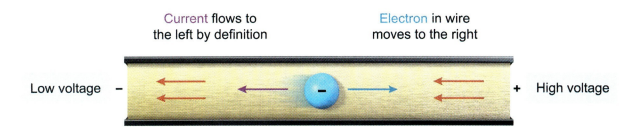

Figure 3.10 Directions of current and electric field, and positions of high and low voltage for an electron moving in a wire.

> ☑ **Concept Check 3.5**
>
> A total of 5.0 × 10¹⁶ electrons pass through a portion of a wire in 2 s. What is the current in the wire?
>
> **Solution**
>
> *Note: The appendix contains the answer.*

3.2.02 Conductivity and Resistivity

The ability of charge to move through a material depends both on the intrinsic properties of the material and the material's size and shape. **Conductivity** σ is an intrinsic property that describes the ease with which charges flow through the molecular structure of a given material. Conversely, **resistivity** ρ measures the degree to which the material impedes the flow of charge.

Conductivity is inversely related to resistivity; hence, a material that has a high conductivity is also characterized by low resistivity (and vice versa):

$$\sigma = \frac{1}{\rho}$$

Materials can be broadly classified as either conductors or insulators. **Conductors** allow the flow of charge when they are exposed to a voltage, typically because the electrons in the atoms of conductors are mobile. Microscopically, conductors and insulators differ in that the outermost (ie, valence) electrons of conductors can detach and move freely throughout the conductor. However, **insulators** do not contain these "conduction electrons" (ie, mobile charge carriers), and thus do not allow charge to flow when a voltage is applied (see Figure 3.11).

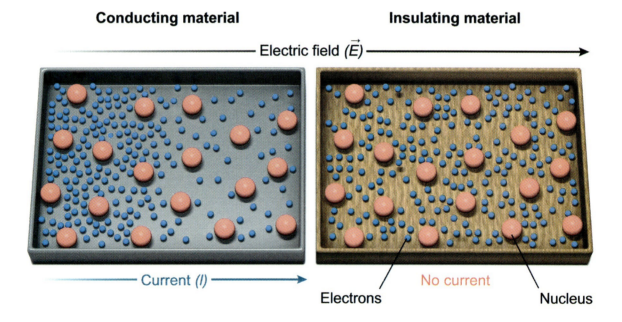

Figure 3.11 Electrons in conducting and insulating materials.

Most **metals** are highly conductive whereas most **nonmetals** are nonconductive or only partially conductive. Metals are highly conductive because valence electrons within metals are loosely associated with each atom's nuclei. Hence, the application of an electric field to a metal easily dislodges valence electrons and generates a current.

In contrast to solid objects like metals and nonmetals, electrolytic solutions may be conductive in the absence of freely moving valence electrons. Instead, the conductivity of electrolytic solutions is most closely related to the concentration of ions because the motion of these mobile charge carriers throughout the fluid generates current. Figure 3.12 compares the conductivity and resistivity for various substances.

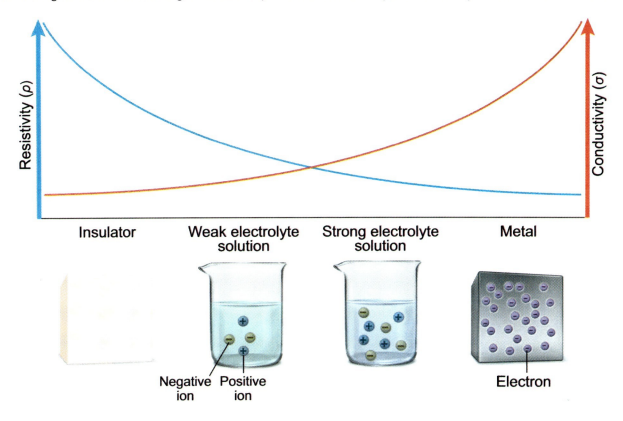

Figure 3.12 Conductivity and resistivity for various substances.

The total conductance and **resistance** of a material also depends on the size of the material sample. The sample resistance R is defined as the product of the resistivity and the ratio of the sample's length L to its cross-sectional area A (Figure 3.13):

$$R = \rho \frac{L}{A}$$

For a sample of wire, the resistance is directly proportional to the wire's length but inversely proportional to its cross-sectional area. The unit for resistance is the **ohm** (Ω).

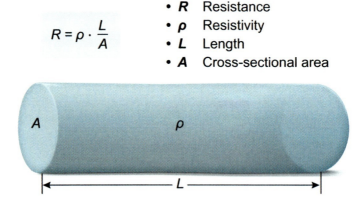

Figure 3.13 Resistance of a piece of wire.

For example, suppose a cylindrical sample of metal has a resistance of 100 Ω, a length of 0.3 m, and a cross-sectional area of 0.1 m². The equation for R can be rearranged to determine the resistivity:

$$R = \rho \frac{L}{A} \Rightarrow \rho = R \frac{A}{L}$$

$$\rho = 100 \, \Omega \left(\frac{0.1 \text{ m}^2}{0.3 \text{ m}} \right)$$

$$\rho = 33 \, \Omega \cdot \text{m}$$

 Concept Check 3.6

A researcher wants to reduce the resistance of a sample of wire by a factor of four. How could the researcher change the size of the wire to accomplish this?

Solution

Note: The appendix contains the answer.

3.2.03 Circuit Basics: Resistance and Ohm's Law

The previous section showed how applying a **voltage** across a material generates a **current** that flows with an intrinsic **resistance** provided by the material. The voltage is typically supplied by a battery. In a battery, when a wire is connected from the positive terminal (high voltage) to the negative terminal (low voltage), charge flows from the positive terminal to the negative terminal and creates a basic **circuit** (see Figure 3.14).

In the schematic diagram presented in Figure 3.14, the wire is represented by solid lines emerging perpendicular to the battery terminals. In the representation of the battery, the positive terminal line segment is longer than the negative terminal segment. Wire is designed to have a very low resistance; hence, resistors with given resistance values are added to impede the flow of charge. Resistors are represented schematically by jagged lines.

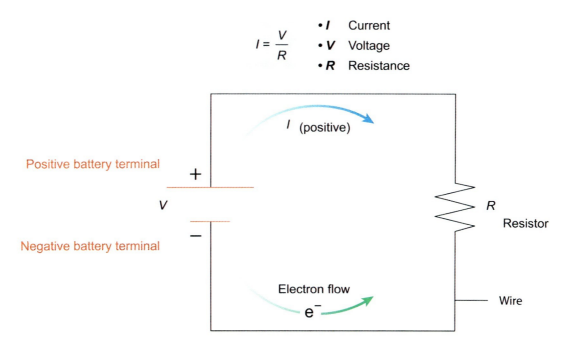

Figure 3.14 Schematic diagram of a basic circuit.

The circulatory system in the body is analogous to a basic circuit. The pumping action of the heart creates a pressure difference that acts as a voltage, pushing blood (ie, the current) through the body. The vascular resistance to the blood flow, which is associated with the viscosity of blood (Concepts 2.2.04 and 2.2.05), is analogous to the resistance of a resistor (Figure 3.15)

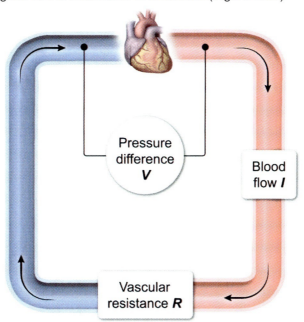

Figure 3.15 Circulatory system as a basic circuit.

Ohm's law states that the amount of current I is equal to the ratio of the voltage applied V and the resistance R:

$$I = \frac{V}{R}$$

Consequently, current is directly proportional to the applied voltage but inversely proportional to the resistance. Note that Ohm's law is analogous to Poiseuille's law of viscous flow, with I corresponding to the volumetric flow rate Q and V to the pressure difference ΔP:

$$I = \frac{V}{R} \quad \Rightarrow \quad Q = \frac{\Delta P}{\left(\frac{8\eta L}{\pi r^4}\right)}$$

Circuit resistance is analogous to a combination of viscosity η and the dimensions of the conduit.

Ohm's law is crucial for analyzing circuits and determining unknown variables. For example, suppose that a 12 V battery creates a current of 4 A in the basic circuit shown in Figure 3.16.

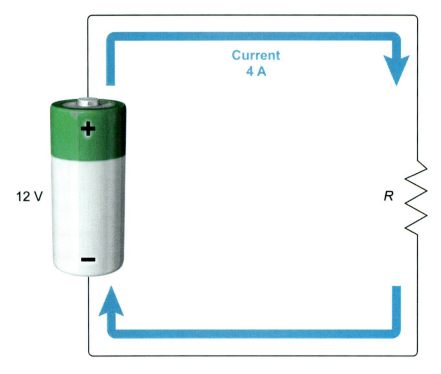

Figure 3.16 Applying Ohm's law to determine resistance given voltage and current.

Inserting these given values into Ohm's law results in:

$$4\text{ A} = \frac{12\text{ V}}{R}$$

Rearranging this equation to solve for R yields:

$$R = \frac{12\text{ V}}{4\text{ A}} = 3\ \Omega$$

 Concept Check 3.7

A group of students measures a current of 5 A passing through a 3 Ω resistor in a basic circuit. What is the voltage of the battery in the circuit?

Solution

Note: The appendix contains the answer.

As a charge Q moves from high to low voltage in a circuit, the change in **electric potential energy** ΔE is equal to the product of Q and V (Concept 3.1.04):

$$\Delta E = QV$$

Furthermore, **power** P (Concept 1.5.05) is the change in energy per unit time. Hence, in an electric circuit, P is the product of I and V:

$$P = \frac{\Delta E}{t} = \left(\frac{Q}{t}\right)V$$

$$P = IV$$

The formula for power can be used to calculate the rate that energy is dissipated by an electric circuit. For example, resistors dissipate electric energy in the form of heat (Chapter 5), and light bulbs dissipate electric energy both in heat and emitted light (see Figure 3.17). Using Ohm's law, voltage can be eliminated to express P in terms of only I and R:

$$P = I(IR) = I^2 R$$

Similarly, Ohm's law can also be used to eliminate current, leading to an expression for P in terms of only V and R:

$$P = \left(\frac{V}{R}\right)V = \frac{V^2}{R}$$

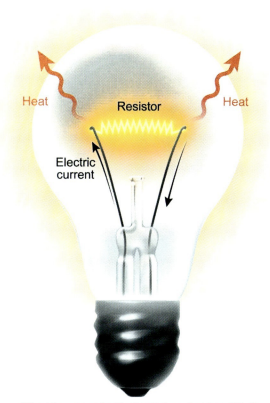

Figure 3.17 Electric power is dissipated as heat and light.

> ## ☑ Concept Check 3.8
>
> A basic circuit contains a 10 V battery and a resistor of 5 Ω. What is the power dissipated by the resistor?
>
> ### Solution
>
> *Note: The appendix contains the answer.*

In the real world, the electric energy per unit time supplied by a battery to move charge (known as the **electromotive force** $\mathcal{E}$, or emf) is not equal to the actual voltage across the terminals (i.e., the **terminal voltage** V_{term}) observed when the battery is connected to a circuit. The reason for this discrepancy is that the battery has an **internal resistance** r that dissipates energy as a current I flows through the circuit.

Ohm's law implies that the voltage lost V_{loss} due to internal resistance is equal to the product of I and r:

$$V_{loss} = Ir$$

Consequently, the discrepancy between electromotive force and the observed terminal voltage is:

$$V_{term} = \mathcal{E} - V_{loss} = \mathcal{E} - Ir$$

A battery with internal resistance can be represented schematically in a circuit, as shown in Figure 3.18.

$$V_{term} = \mathcal{E} - Ir = IR$$

- V_{term} Potential difference between battery terminals
- $\mathcal{E}$ Electromotive force
- I Current
- r Internal resistance
- R Circuit resistance

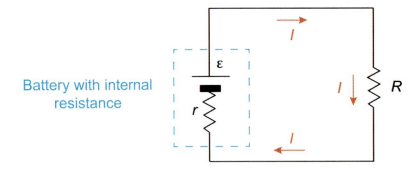

Figure 3.18 A basic circuit containing a battery with an internal resistance.

3.2.04 Series and Parallel Circuits

Basic circuits were introduced in the previous section, where a battery is connected to one resistor using a wire that runs from the positive terminal to the negative terminal. However, more complicated circuits that include more than one component connected to the battery are often encountered.

Circuit components are defined as connected either in series or in parallel, depending on their arrangement. **Series circuits** position components in linear succession and form a single path while a fixed quantity of current passes through each element (Figure 3.19).

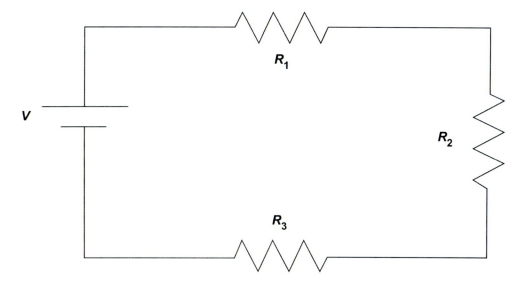

Figure 3.19 Schematic diagram of a series circuit containing three resistors (R_1, R_2, R_3).

In contrast, **parallel circuits** position components with multiple pathways through which current can flow. The branching and merging of current flow across multiple channels occurs at junction points (Figure 3.20).

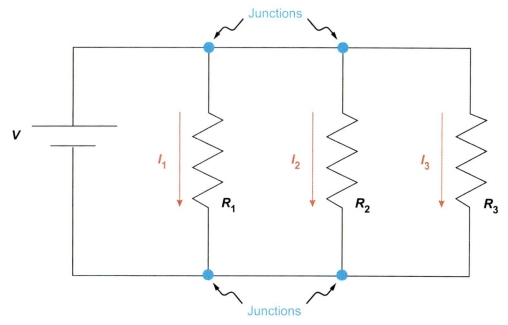

Figure 3.20 Schematic diagram of three resistors in parallel.

The behavior of the current through and the voltages across each resistor can be determined by Kirchhoff's rules (ie, the junction rule and the loop rule). **Kirchhoff's junction rule**, which follows from the conservation of charge, states that the total current I_{enter} entering a circuit junction must equal the total current I_{exit} exiting the junction (see Figure 3.21):

$$I_{enter} = I_{exit}$$

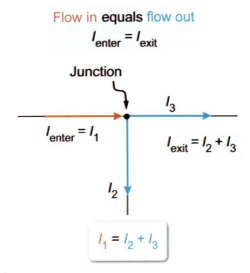

Figure 3.21 Kirchhoff's junction rule.

Electric circuits also exhibit conservation of energy around a closed path (loop) in a circuit, in which the direction of the path matches the direction of the current in the loop. As a result, **Kirchhoff's loop rule** states that the sum of the voltage V across components around any closed loop in a circuit equals zero:

$$\sum_{i=1}^{n} V_i = V_1 + V_2 + \cdots V_n = 0$$

Consequently, in a series circuit the current I is the same through all resistors, and the sum of the voltage V_i across each resistor equals the battery voltage V (Figure 3.22).

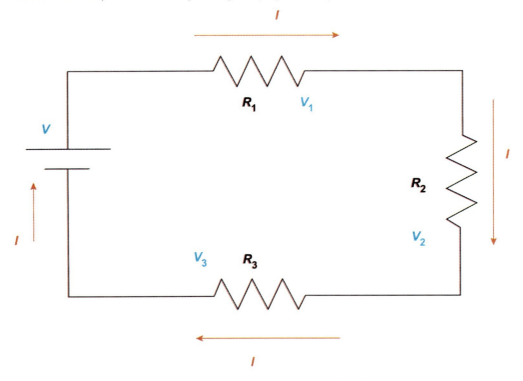

Figure 3.22 Current and voltage in a series circuit.

In a parallel circuit, the sum of the currents in each branch I_i equals the total current I through the battery, and the voltage across each branch of the circuit equals the battery voltage V (Figure 3.23):

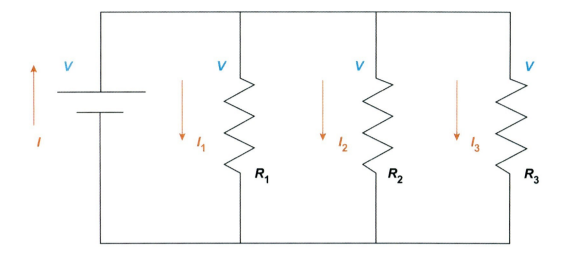

- Same **voltage** through battery and resistors
- Sum of currents in each branch I_i equals total current

$$I = I_1 + I_2 + I_3$$

Figure 3.23 Current and voltage in a parallel circuit.

Furthermore, Ohm's law implies that the voltages across each resistor are equal to the product of the currents and resistances:

$$V_i = I_i \cdot R_i$$

Using Ohm's law, it is possible to derive a **total equivalent resistance** R_{eq} for each type of circuit. The R_{eq} for resistors connected in series is equal to the sum of the individual resistors:

$$R_{eq} = R_1 + R_2 + R_3 + \cdots$$

Here R_{eq} is always *greater than* the resistance of any of the individual resistors. On the other hand, the inverse of the R_{eq} for resistors connected in parallel is equal to the sum of the inverses of the individual resistors:

$$\frac{1}{R_{eq}} = \frac{1}{R_1} + \frac{1}{R_2} + \frac{1}{R_3} + \cdots$$

For resistors connected in parallel, R_{eq} is always *less than* the resistance of any of the individual resistors.

These two equations are important to remember as they are often applied to circuit problems on the exam.

Concept Check 3.9

Three resistors are connected to a battery of voltage V, as shown in the circuit below.

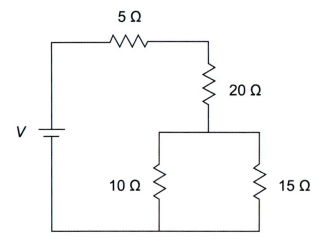

What is the total equivalent resistance of the circuit?

Solution

Note: The appendix contains the answer.

Concept Check 3.10

Two resistors, R_1 and R_2, are connected in parallel to a 12 V battery.

(a) What is the voltage across R_1?

(b) What is the voltage across R_2?

(c) Suppose that a current of 3 A passes through R_1 and a current of 2 A passes through R_2. What is the current passing through the battery?

Resistor R_1 is removed from the circuit, leaving R_2 connected in series with the battery.

(d) How does the total equivalent resistance of the battery change after R_1 is removed?

(e) How does the current through R_2 change?

Solution

Note: The appendix contains the answer.

3.2.05 Meters

The behavior of electrical devices depends crucially on the values of current, voltage, and resistance. An **ammeter** and a **voltmeter** are used to measure the current and voltage, respectively, through an element in a circuit. An **ohmmeter** is used to measure resistance in a circuit; however, the details of ohmmeter design and function are not typically covered on the exam.

To properly measure the current, an ammeter must be placed *in series* with a resistor (see Figure 3.24).

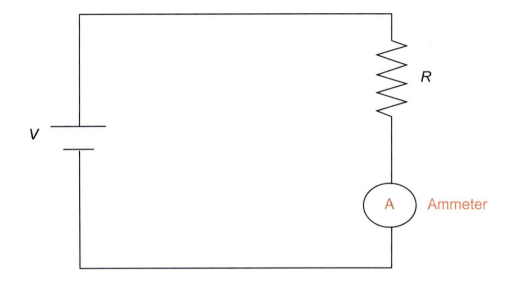

Figure 3.24 Schematic diagram of an ammeter connected in series with a resistor.

Placing the ammeter in the circuit alters the current in the circuit. Initially, the current $I_{initial}$ through the resistor R is equal to the battery voltage V divided by R:

$$I_{initial} = \frac{V}{R}$$

However, with the ammeter in the circuit the final current equals:

$$I_{final} = \frac{V}{R + R_{meter}}$$

where R_{meter} is the resistance of the ammeter. Consequently, an ammeter must have a *low resistance* (ie, R_{meter} must be much smaller than R) for the meter to give an accurate reading:

$$R_{meter} \ll R \implies I_{final} \approx I_{initial}$$

In contrast, a voltmeter should be placed *in parallel* with a resistor to measure the voltage across the resistor (see Figure 3.25).

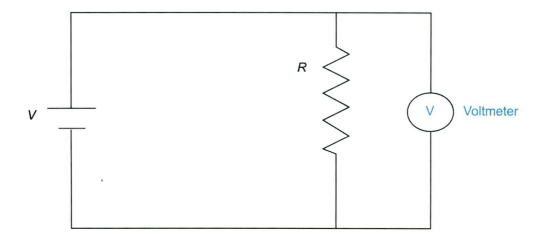

Figure 3.25 Schematic diagram of a voltmeter connected in parallel with a resistor.

The addition of the voltmeter decreases the total effective resistance of the circuit R_{eq} according to:

$$\frac{1}{R_{eq}} = \frac{1}{R} + \frac{1}{R_{meter}}$$

Ohm's law implies that a decrease in the total resistance increases the total current through the circuit, affecting the voltage across R. Hence, a voltmeter should have a *high resistance* (ie, R_{meter} must be much greater than R).

$$R_{meter} \gg R \;\Rightarrow\; V_{final} \approx V_{initial}$$

✓ Concept Check 3.11

How should an ammeter and a voltmeter be placed in the circuit below to correctly measure the current and voltage through resistor R_2?

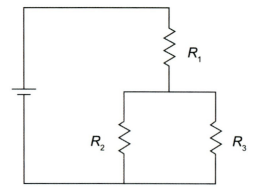

Solution
Note: The appendix contains the answer.

Lesson 3.3
Capacitance

Introduction

Capacitors are the third type of circuit element on the exam, along with batteries and resistors. Capacitors can store electrical energy by separating positive and negative charges. In particular, the exam focuses on the electrical characteristics and physical properties of parallel plate capacitors.

3.3.01 Definition of Capacitance

A simple **parallel plate capacitor** consists of two flat, metallic plates evenly separated by a gap, as shown in Figure 3.26. Connecting a capacitor to a battery charges the capacitor by storing electric charge on the plates. When the capacitor is charging, electric current flows out of the negatively charged plate and into the positively charged plate until the capacitor voltage is equal to the battery voltage.

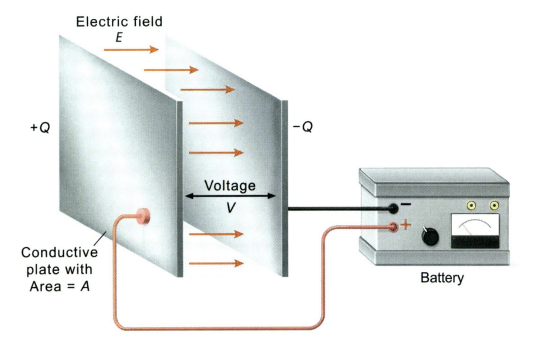

Figure 3.26 Charge and voltage stored in a simple parallel plate capacitor.

The equal but opposite charges stored on the capacitor plates generate a voltage due to the separation of the positive and negative charges. The **capacitance** C of a parallel plate capacitor depends on its **physical characteristics** and is equal to the ratio of the stored electric charge Q and the voltage V:

$$C = \frac{Q}{V}$$

The units for capacitance are farads (F), where 1 F equals 1 coulomb (C) of charge stored per volt (V) of electric potential:

$$1\,\text{F} = 1\,\frac{\text{C}}{\text{V}}$$

A **real-life capacitor** can only store a limited amount of electric charge on its plates. When a large amount of positive charge accumulates on the positively charged plate, electrostatic forces prevent the addition of more positive charge. This limitation for practical capacitors is often referred to as the **voltage limit** for a specific device.

The symbol for a capacitor in a **circuit diagram** is two parallel lines to represent the parallel plates (see Figure 3.27).

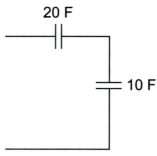

Figure 3.27 Circuit diagram containing a 20 F capacitor and a 10 F capacitor.

☑ **Concept Check 3.12**

A 1 mF capacitor has a voltage limit of 50 V. What is the maximum amount of charge the capacitor can store on its positive plate?

Solution

Note: The appendix contains the answer.

3.3.02 Energy of a Charged Capacitor

When electric charges are stored on the plates of a capacitor, an electric field is generated between the plates due to the separation of the positive and negative charges. The capacitor stores **electric energy** in this electric field, as shown in Figure 3.28.

The energy U stored in the capacitor is equal to half the product of the capacitance C and the square of the voltage V across the plates:

$$U = \frac{1}{2}CV^2$$

Thus, U is directly proportional to C and proportional to the square of V. As with work and other forms of energy, the units for U are joules (J).

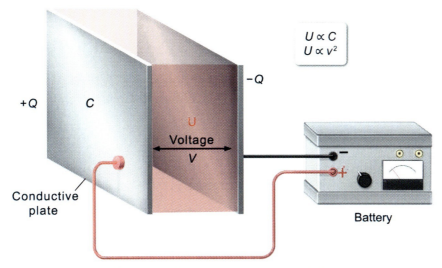

Figure 3.28 Energy stored in a capacitor depends on the capacitance and the voltage.

Recall that C is equal to the ratio of charge Q and V:

$$C = \frac{Q}{V}$$

Substituting this relationship into the equation for U gives an alternative method for calculating U as half the product of Q and V:

$$U = \frac{1}{2}CV^2 = \frac{1}{2}\left(\frac{Q}{\cancel{V}}\right)V^{\cancel{2}} = \frac{1}{2}QV$$

$$U = \frac{1}{2}QV$$

Furthermore, V can be calculated as the ratio of Q and C:

$$V = \frac{Q}{C}$$

This relationship yields a third equation that shows U is equal to half the ratio of Q squared and C:

$$U = \frac{1}{2}CV^2 = \frac{1}{2}C\left(\frac{Q}{C}\right)^2 = \frac{1}{2}\frac{\cancel{C}Q^2}{C^{\cancel{2}}} = \frac{1}{2}\frac{Q^2}{C}$$

$$U = \frac{1}{2}\frac{Q^2}{C}$$

 Concept Check 3.13

How much energy is required to increase the voltage across a 0.2 F capacitor from 1 V to 10 V?

Solution

Note: The appendix contains the answer.

3.3.03 Parallel Plate Capacitors

The **capacitance** of a parallel plate capacitor depends on its **physical characteristics**, including the **geometry** of the plates. Because the voltage V is equal to the product of electric field strength E and the distance d between the plates, capacitance C is inversely proportional to d:

$$C = \frac{Q}{V} = \frac{Q}{E \cdot d}$$

$$C \propto \frac{1}{d}$$

As shown in Figure 3.29, E is proportional to the ratio of the charge Q and the area A of the plates:

$$E \propto \frac{Q}{A}$$

Increasing the plate area in Figure 3.29 causes the electric field lines to be spaced further apart. As discussed in Concept 3.1.03, the magnitude of an electric field is larger when the electric field lines are closer together. As a result, increasing the plate area reduces the electric field strength and increases the spacing between the electric field lines.

Because C is inversely proportional to E, C is also directly proportional to A:

$$C \propto \frac{1}{E} \propto A$$

Combining these relationships shows C is dependent on the ratio of A and d:

$$C \propto \frac{A}{d}$$

$$E \propto \frac{Q}{A}$$

$$C \propto \frac{1}{E} \propto \frac{A}{Q}$$

$$C \propto A$$

- **E** Electric field
- **Q** Charge
- **A** Area
- **C** Capacitance

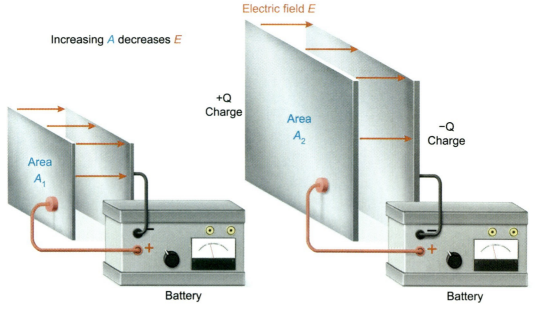

Figure 3.29 Increasing plate area decreases electric field and increases capacitance.

Another important physical characteristic that influences the capacitance of a parallel plate capacitor is the electrical insulating properties of the material separating the plates. This material's insulation ability depends on its **dielectric constant** κ, which is a unitless quantity with a value greater than or equal to 1 (Table 3.1).

Table 3.1 Dielectric constants of various materials.

Material	Dielectric constant (κ)
Vacuum	1.0
Teflon	2.1
Rubber	3.0
Paper	3.5
Silicon	10

A **vacuum** has a κ equal to 1, whereas capacitors use dielectrics with κ values greater than 1. Figure 3.30 shows that a capacitor with a vacuum between the plates has a C equal to C_0. If a dielectric with a value of κ is used instead of a vacuum, the capacitor's C increases by a factor of κ:

$$C = \kappa C_0$$

This occurs because materials with a larger κ are better insulators, which reduces the voltage between the capacitor plates for the same amount of electric charge on the plates.

Capacitance C is equal to the ratio of charge Q to voltage V:

$$C = \frac{Q}{V}$$

Inserting a dielectric material between the capacitor plates decreases the value of V while Q remains constant, causing the value of C to increase:

$$\frac{Q}{\downarrow V} = \uparrow C$$

From Concept 3.1.04, V is equal to the product of electric field strength E and the distance d between the plates:

$$V = E \cdot d$$

Consequently, decreasing V also causes E to decrease because d is constant:

$$\downarrow V = \downarrow E \cdot d$$

Therefore, capacitors using a material with a larger κ have a greater C, lower V, and lower E (assuming identical Q, A, and d).

$C = \kappa \cdot C_0$

- C Capacitance with dielectric
- κ Dielectric constant
- C_0 Capacitance of a vacuum

Vacuum between the plates
$\kappa = 1.0$

$C = C_0$

Higher voltage
Larger electric field

Dielectric between the plates
$\kappa = 1.2$

$C = 1.2 \cdot C_0$

Lower voltage
Smaller electric field

Figure 3.30 A capacitor with a dielectric between the plates increases the capacitance by a factor of κ.

Furthermore, C can be directly calculated from κ, d, A, and the **permittivity of free space** ε_0 using the relationship:

$$C = \kappa \varepsilon_0 \frac{A}{d}$$

where ε_0 is a constant equal to 8.8×10^{-12} F/m. Note that ε_0 is related to the Coulomb constant k discussed in Concept 3.1.02 via the relation:

$$k = \frac{1}{4\pi\varepsilon_0}$$

Concept Check 3.14

A capacitor with a plate separation of 1 mm and plate area of 1 cm² has a capacitance of 5 μF. What is the capacitance if the plate separation is changed to 0.5 mm and the plate area is changed to 5 cm²?

Solution

Note: The appendix contains the answer.

3.3.04 Capacitors in Series and Parallel

Complex circuits may contain **multiple capacitors** connected in series or in parallel. Like circuits with multiple resistors (Concept 3.2.04), a circuit with multiple capacitors can be simplified to a single **equivalent capacitance**, although the equations for combining capacitors are different than those for resistors.

Capacitors connected in **parallel** (Figure 3.31) can be replaced with a single equivalent capacitance C_{Eq} by summing the individual capacitor values:

$$C_{Eq} = C_1 + C_2 + C_3 + \cdots$$

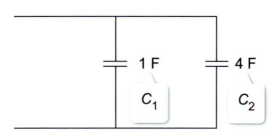

Figure 3.31 Capacitors connected in parallel.

When capacitors are connected in parallel, C_{Eq} is always *greater than* the capacitance of any of the individual capacitors. For example, C_{Eq} for the circuit shown in Figure 3.31 is given by:

$$C_{Eq} = 1\text{ F} + 4\text{ F} = 5\text{ F}$$

Alternatively, capacitors connected in **series** (Figure 3.32) can be replaced with a single C_{Eq} using the equation:

$$\frac{1}{C_{Eq}} = \frac{1}{C_1} + \frac{1}{C_2} + \frac{1}{C_3} + \cdots$$

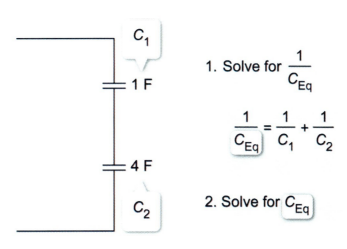

Figure 3.32 Capacitors connected in series.

When capacitors are connected in series, C_{Eq} is always *less than* the capacitance of any of the individual capacitors. For example, C_{Eq} for the circuit shown in Figure 3.32 is given by:

$$\frac{1}{C_{Eq}} = \frac{1}{1\text{ F}} + \frac{1}{4\text{ F}} = \frac{4}{4\text{ F}} + \frac{1}{4\text{ F}} = \frac{5}{4\text{ F}}$$

$$C_{Eq} = \frac{4\text{ F}}{5} = 0.8\text{ F}$$

Circuits containing multiple capacitors connected in series and in parallel are simplified by calculating C_{Eq} for each combination of capacitors until a final capacitance is reached. For example, C_{Eq} of the circuit shown in Figure 3.33 is calculated by combining the two 2 F capacitors connected in parallel and then combining that equivalent capacitor in series with the 4 F capacitor.

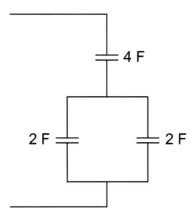

Figure 3.33 Complex circuit with three capacitors.

The parallel combination of the two 2 F capacitors yields a C_{Eq} of 4 F:

$$C_{Eq} = 2\text{ F} + 2\text{ F} = 4\text{ F}$$

Figure 3.34 shows the simplified circuit consisting of only two 4 F capacitors in series.

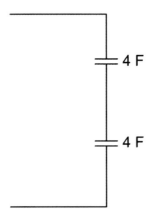

Figure 3.34 Simplified circuit after combining parallel capacitors.

The series combination of these 4 F capacitors yields a further C_{Eq} of 2 F:

$$\frac{1}{C_{Eq}} = \frac{1}{4\text{ F}} + \frac{1}{4\text{ F}} = \frac{2}{4\text{ F}}$$

$$C_{Eq} = \frac{4\text{ F}}{2} = 2\text{ F}$$

Figure 3.35 shows the final simplified circuit diagram containing only a single 2 F capacitor.

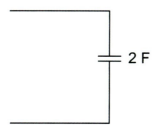

Figure 3.35 Simplified circuit after combining series capacitors.

Concept Check 3.15

A researcher wants to combine two 0.5 mF capacitors to make an equivalent capacitor with the greatest possible capacitance. Should the two capacitors be combined in series or in parallel to make the greatest possible equivalent capacitance?

Solution

Note: The appendix contains the answer.

Lesson 3.4
Magnetism

Introduction

This lesson describes the magnetic field, which is created by moving charges and is complementary to the electric field. Electric fields and magnetic fields are different components of one overall electromagnetic field. Magnetic fields are commonly associated to permanent magnets, which have north and south poles.

Charges moving in a magnetic field experience a Lorentz force perpendicular to the direction of their motion. Consequently, the magnetic field deflects the charge, and the charge exhibits circular motion while moving within the field.

3.4.01 Magnets and the Magnetic Field

When a charge (eg, proton, electron) or charged object is in motion, the electric field in the surrounding space changes. Consequently, this motion generates what is known as a **magnetic field** around the moving charge.

For example, at the subatomic level the movement of electrons around the nuclei of certain atoms can generate stable magnetic fields. Materials in which a magnetic field is present are known as **magnets**. Common sources of magnetic fields include electromagnetic radiation and **permanent magnets**, materials in which sustained north and south magnetic poles are present intrinsically or formed through magnetic induction (ie, magnetization).

Magnets always have two poles (ie, north and south), even when split into pieces. As such, a single pole of a magnet cannot exist, unlike electric charges in which a positive and negative charge can exist independently (Figure 3.36).

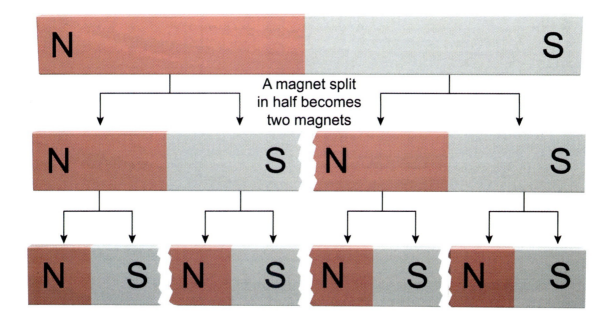

Figure 3.36 A magnetic north and south pole exists on all magnets, even when broken into pieces.

Just like with electric charges, similar poles on different magnets always repel one another whereas opposite poles always attract (Figure 3.37).

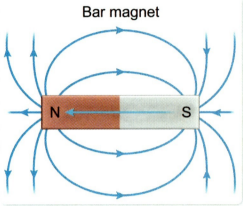

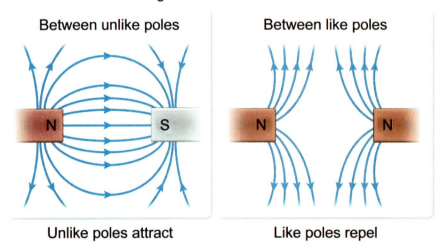

Figure 3.37 Magnetic field lines generated by permanent magnets.

The resulting magnetic field generated by magnetized objects or materials is a vector quantity, with units of teslas (T) and represented in diagrams by **magnetic field lines**. The field itself represents the magnetic influence that other *moving* charges (ie, electric currents) and magnetized materials would experience at a specific location in three-dimensional space.

The strength of the magnetic field on a diagram is depicted by the density of field lines (ie, the distance between the lines), and the direction of a field line represents the direction that a compass needle would point if placed on the line itself, as shown in Figure 3.38.

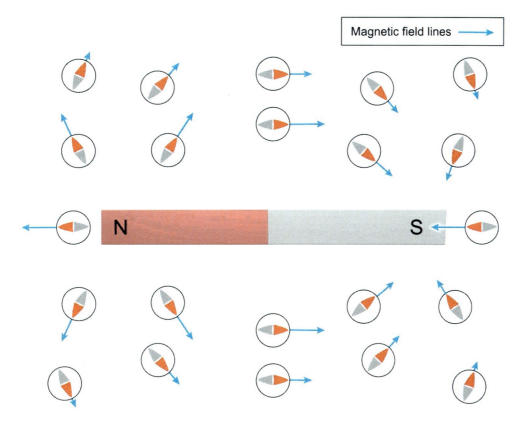

Figure 3.38 A compass aligns itself parallel to invisible magnetic field lines and toward south magnetic pole.

On Earth, this can be somewhat misleading, as a compass will align itself toward the north *geographic* pole because that is where the south *magnetic* pole (see Figure 3.39) is located.

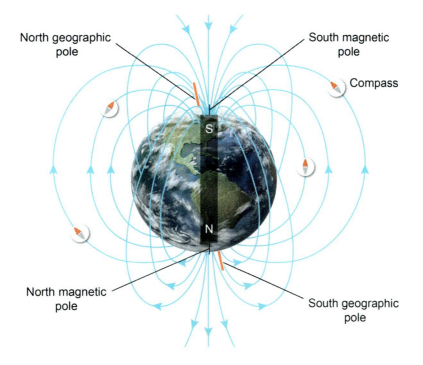

Figure 3.39 The south magnetic pole of Earth is located near the north geographic pole.

Thus, magnetic field lines are continuous, never cross, and form closed loops that exit from the north pole of a magnet and enter at the south pole.

> ☑ **Concept Check 3.16**
>
> In which of the following situations is a magnetic field generated in the surrounding space? There may be more than one correct answer.
>
> A. An electron is at rest
> B. A proton and an electron are held at rest a fixed distance apart
> C. A current flows through a wire
> D. A proton and an electron accelerate toward one another
>
> **Solution**
>
> Note: The appendix contains the answer.

3.4.02 Lorentz Force

The **Lorentz force** is the magnetic force that acts on a charge *moving* within a magnetic field. The Lorentz force F, the velocity v of the charge, and the magnetic field B are all vectors with both magnitude and direction. The magnitude of F on a charge q depends on the angle θ between the directions of v and B according to the equation:

$$F = qvB \sin \theta$$

Moreover, by rearranging the Lorentz force equation and isolating B, the dimensions representing the unit of tesla (T) can be derived:

$$B = \frac{F}{qv \sin \theta}$$

$$1 \text{ T} = 1 \, \frac{\text{N}}{\text{C} \cdot \frac{\text{m}}{\text{s}}}$$

The ampere (A) can be placed into the denominator to replace coulomb per second (C/s):

$$1 \text{ T} = 1 \, \frac{\text{N}}{\frac{\text{C}}{\text{s}} \cdot \text{m}} = 1 \, \frac{\text{N}}{\text{A} \cdot \text{m}}$$

Thus, the SI units representing the tesla are newtons per ampere-meter (N/A·m).

The direction of F is always perpendicular to both the direction of v and the direction of B (see Figure 3.40).

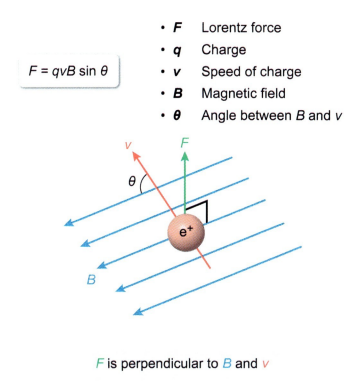

$F = qvB \sin \theta$

- **F** Lorentz force
- **q** Charge
- **v** Speed of charge
- **B** Magnetic field
- **θ** Angle between B and v

F is perpendicular to B and v

Figure 3.40 Magnetic force on a charge moving in a magnetic field.

The exact orientation of F, B, or v can be determined by using the **right-hand rule** when two of the three vector directions are known. By forming a backward letter "L" with the right thumb and index finger, and lowering the middle finger to the horizontal, the three perpendicular axes representing the three vectors can be visualized. The thumb represents F, the index finger represents v, and the middle finger represents B, as shown in Figure 3.41. It is important to note that the direction of F represents the force on a *positive charge*, and the force on a negative charge is *directed opposite to the direction of the thumb*.

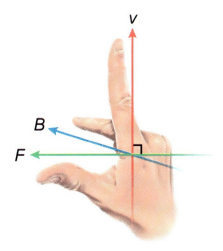

Figure 3.41 The direction of the Lorentz force on a moving positive charge with speed v in a magnetic field B.

Because the magnetic force on a charge is always perpendicular to the velocity vector (ie, the direction of motion), the charge subsequently moves along a circular path with a centripetal acceleration (*a*), which is directed radially inward and equal to the ratio of the velocity *v* squared and the radius of rotation *r*:

$$a = \frac{v^2}{r}$$

Thus, the Lorentz force is a centripetal force, and the charge rotates about a magnetic field line with a constant period (Figure 3.42).

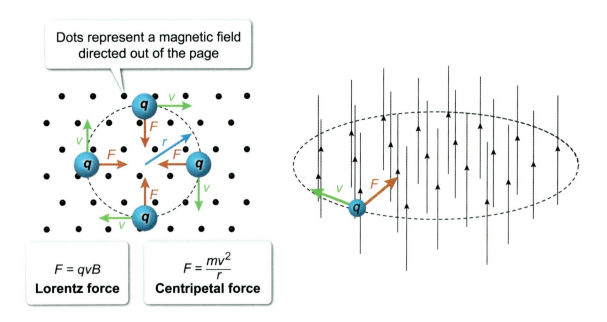

Figure 3.42 A charge moving in a magnetic field directed out of the page experiences a centripetal Lorentz force.

Although a charge moving in a magnetic field experiences a Lorentz force, a moving charge also generates a magnetic field of its own. For example, a wire carrying a current *I* (ie, a set of moving charges) generates a magnetic field around the wire that acts in a circular direction.

To determine the direction of the magnetic field generated by a current in a wire, another variation of the right-hand rule can be useful. With the thumb directed parallel to the current direction, the fingers can be rotated (ie, curled) around the imaginary wire. The rotational direction of the fingers represents the direction of the magnetic field around the wire (see Figure 3.43).

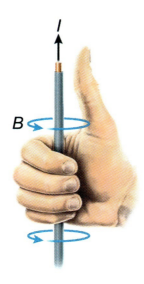

Figure 3.43 The direction of the magnetic field generated by a current flowing through a wire.

 Concept Check 3.17

An electron moves near Earth's surface in an eastward direction where the magnetic field is directed northward. What is the direction of the Lorentz force experienced by the electron: northward, southward, eastward, westward, toward the ground, or away from the ground?

Solution

Note: The appendix contains the answer.

 Concept Check 3.18

A positron of charge $+e$ and mass m moves in a magnetic field B with speed v in uniform circular motion. What is the positron's radius of rotation in terms of m, B, e, and v?

Solution

Note: The appendix contains the answer.

END-OF-UNIT MCAT PRACTICE

Congratulations on completing **Unit 3: Electrostatics and Circuits**.

Now you are ready to dive into MCAT-level practice tests. At UWorld, we believe students will be fully prepared to ace the MCAT when they practice with high-quality questions in a realistic testing environment.

The UWorld Qbank will test you on questions that are fully representative of the AAMC MCAT syllabus. In addition, our MCAT-like questions are accompanied by in-depth explanations with exceptional visual aids that will help you better retain difficult MCAT concepts.

TO START YOUR MCAT PRACTICE, PROCEED AS FOLLOWS:

1) Sign up to purchase the UWorld MCAT Qbank
   IMPORTANT: You already have access if you purchased a bundled subscription.
2) Log in to your UWorld MCAT account
3) Access the MCAT Qbank section
4) Select this unit in the Qbank
5) Create a custom practice test

Unit 4 Light and Sound

Chapter 4 Waves, Sound, and Light

4.1 Periodic Motion and Waves

    4.1.01 Wave Characteristics
    4.1.02 Longitudinal and Transverse Waves
    4.1.03 Wave Propagation
    4.1.04 Wave Interference

4.2 Sound

    4.2.01 Sound Properties
    4.2.02 Intensity of Sound
    4.2.03 Resonance
    4.2.04 Doppler Effect
    4.2.05 Applications of Sound

4.3 Light

    4.3.01 Light Waves
    4.3.02 Electromagnetic Spectrum
    4.3.03 Light and Matter Interactions
    4.3.04 Reflection and Refraction
    4.3.05 Light Polarization
    4.3.06 Light Interference and Diffraction
    4.3.07 Doppler Effect with Light

4.4 Optical Instruments

    4.4.01 Reflection from Mirrors
    4.4.02 Refraction in Thin Lenses
    4.4.03 Combinations of Lenses and Lens Aberration
    4.4.04 The Human Eye

Lesson 4.1

Periodic Motion and Waves

Introduction

A repeated disturbance in a substance produces a mechanical wave that travels through the material, carrying energy and information about the disturbance. Two types of waves exist, transverse and longitudinal, depending on how the motion of the material compares to the propagation of the wave. Furthermore, the speed of waves depends on the properties of the medium/material/substance and can be expressed in terms of wavelength and frequency.

This lesson begins by introducing the concept of waves and their basic properties of amplitude, wavelength, period, and frequency and concludes by discussing the concept of wave interference, where multiple waves overlap and are combined.

4.1.01 Wave Characteristics

When an **elastic** material is deformed from its natural shape, a **restoring force** pulls the material back to its undeformed shape according to Hooke's law (see Concept 1.3.06). Once released, the deformed elastic material vibrates with a regular repeated motion known as periodic motion, or **oscillation**. The period T of an oscillation is defined as the time it takes to complete one full cycle of motion. Examples of periodic motion include the behavior of a compressed (or stretched) spring and the oscillatory motion of ocean **waves**.

When a vibrating object disturbs a **medium** (eg, air, water, rope), energy from the vibrating object is carried through the medium in the form of a mechanical **wave**. As a wave moves through a medium, the particles of the medium stay essentially fixed to one location as the wave propagates through the medium. For example, as an athlete moves the end of a rope up and down causing the individual pieces of the rope to move up and down, a wave is created that moves along the length of the rope (Figure 4.1).

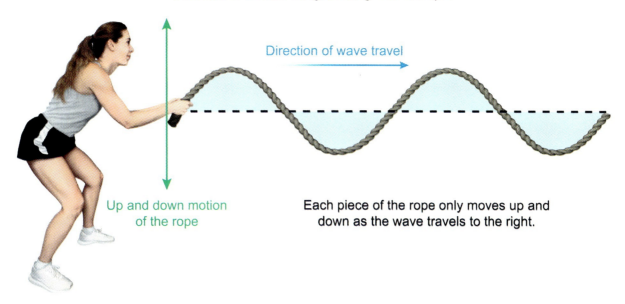

Figure 4.1 Oscillating motion produces waves in a medium.

Waves can be represented mathematically by sine or cosine functions. When plotted on a graph, the *y*-axis portrays the oscillating variable (eg, displacement, pressure, air density) whereas the *x*-axis represents position (Figure 4.2).

The **crest** and **trough** of the wave are the maximum displacement of the medium above and below its undisturbed (ie, equilibrium) position, respectively. The **amplitude** *A* of the wave is the distance from the equilibrium position of the medium to the height of either a crest or a trough.

The **wavelength** λ of a wave is defined as the distance between one point on the wave and the next identical point on the wave. This span is often easy to measure as the distance between two adjacent crests or troughs on the wave.

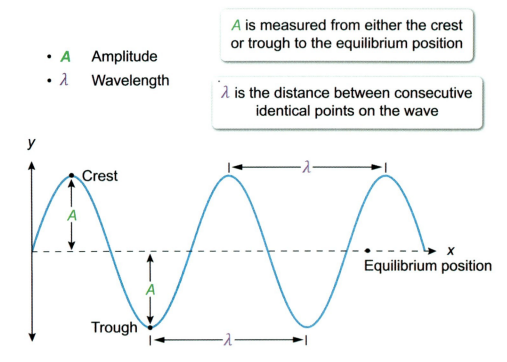

Figure 4.2 A snapshot of a wave at an instant in time.

Plotting the oscillating variable (eg, displacement) as a function of time can also be useful. The **period** *T* of an oscillation is defined as the time it takes for one complete cycle. For example, the period represents the time it takes for each piece of the rope in Figure 4.1 to move up and back down in one complete cycle of motion. As a result, the period of a wave is equal to the time for one wavelength to pass a fixed point in the medium, as shown in Figure 4.3. Thus, *T* is measured in units of seconds (s) per wavelength.

The **frequency** *f* is defined as the number of waves that pass a certain point in 1 s, or wavelengths per second. Thus, frequency is the *inverse* of period:

$$f = \frac{1}{T}$$

Since *T* is measured in units of seconds, frequency is measured in hertz (Hz), where:

$$1\,\text{s}^{-1} = 1\,\text{Hz}$$

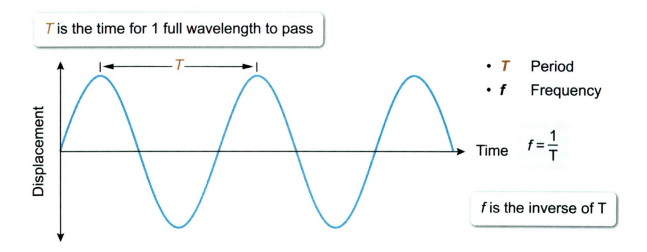

Figure 4.3 The period and frequency of a wave.

For example, suppose a wave travels through water across the surface of a lake. One wavelength takes 2 s to pass the post of a dock. The period T of the wave is equal to 2 s, and the frequency f of the wave is equal to:

$$f = \frac{1}{T} = \frac{1}{(2\text{ s})} = 0.5 \text{ Hz}$$

✓ Concept Check 4.1

Graphs of the vertical position y as a function of time t and the horizontal position x for a wave moving across the surface of a lake are shown. Determine the values of the amplitude A, wavelength λ, and period T from the graphs and calculate the frequency f of the wave.

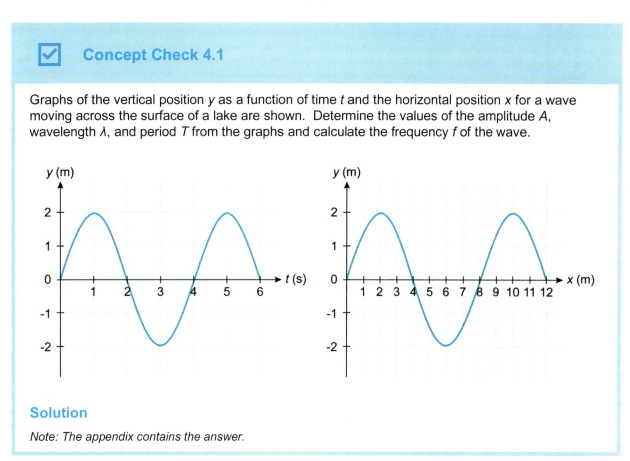

Solution
Note: The appendix contains the answer.

4.1.02 Longitudinal and Transverse Waves

A **transverse wave** is produced when an oscillation displaces the components of the medium *perpendicular* to the direction that the wave travels, as shown in Figure 4.4. Examples of transverse waves include oscillations along a rope, ripples on the surface of water, and electromagnetic waves (which are discussed in detail in Lesson 4.3).

In contrast, a **longitudinal wave** displaces the components of the medium *parallel* to the direction that the wave travels. For example, alternately compressing and stretching one end of a coil spring produces a longitudinal wave (Figure 4.4). The coils move back and forth in a direction *parallel* to the direction of travel of the resulting wave.

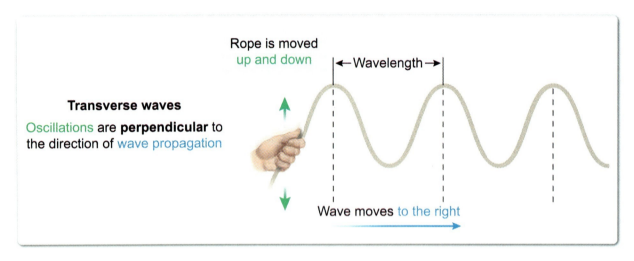

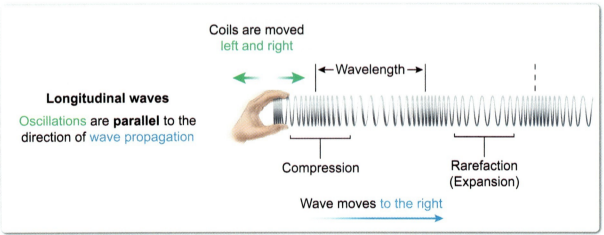

Figure 4.4 Comparing transverse and longitudinal waves.

Longitudinal waves comprise alternating regions of **compression** (where the medium has high density/pressure or contracts) and **rarefaction** (where the medium has low density/pressure or expands). The wavelength of a longitudinal wave can be measured as the distance between the middle of two subsequent compressions or rarefactions, as shown in Figure 4.5.

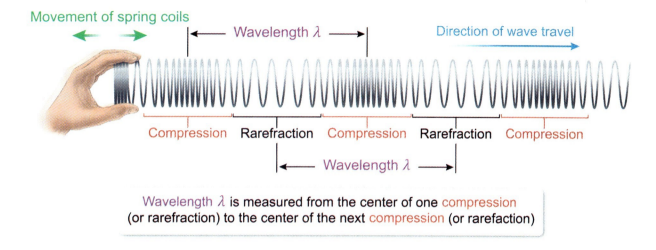

Figure 4.5 Measuring the wavelength of a longitudinal wave.

Perhaps the most common form of longitudinal wave is **sound**. Sound waves are produced by a vibrating object (eg, vocal cords, stereo speaker) that creates alternating regions of high and low pressure in the medium (ie, air, water), as shown in Figure 4.6. The specific properties of sound waves are discussed in Lesson 4.2.

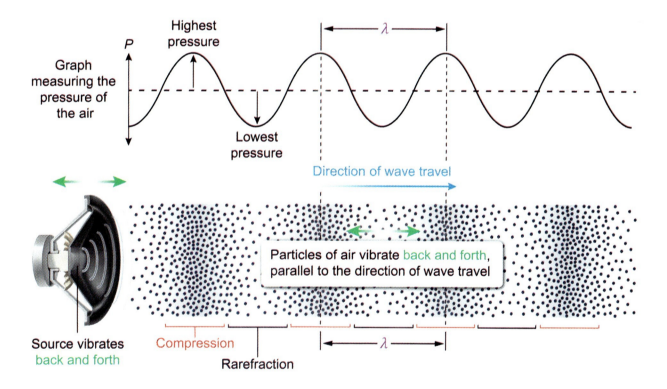

Figure 4.6 Sound as a longitudinal wave.

> ### ✓ Concept Check 4.2
>
> An earthquake produces a seismic transverse wave that propagates horizontally across the surface of the Earth toward a city. In what direction does the ground oscillate as the wave passes?
>
> **Solution**
>
> *Note: The appendix contains the answer.*

4.1.03 Wave Propagation

The previous section described two types of mechanical wave: transverse and longitudinal. Regardless of the type of wave, all mechanical waves **propagate** through a medium with a speed that is independent of how the wave is produced.

Wave speed v is determined by the physical properties of the medium, namely tension and density. Wave speed increases with increased *tension* between particles of the medium and decreases with increased *inertia* of particles of the medium. For example, on the same rope moving up and down, waves increase speed with an increase in tension and decrease speed with an increase in the mass density of the rope.

The speed v of a mechanical wave can be determined from the properties of the wave by measuring the wavelength and either the period or frequency, as shown in Figure 4.7.

$$v = \frac{\lambda}{T} = f\lambda$$

- v Wave speed
- λ Wavelength
- T Period
- f Frequency

$$f = \frac{1}{T}$$

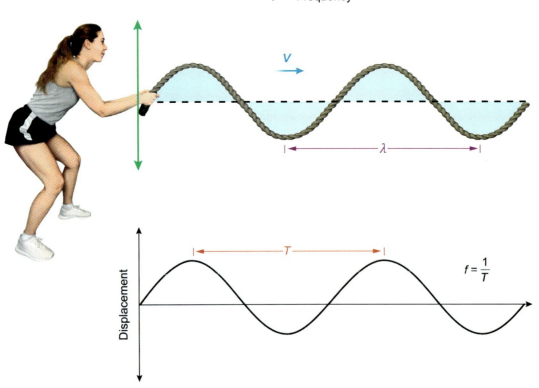

Figure 4.7 Calculating the speed of a wave.

In Concept 1.2.01, speed *v* was defined as distance *d* divided by time *t*:

$$v = \frac{d}{t}$$

As described earlier, one period *T* is equal to the time for one wavelength *λ* of a wave to pass a fixed point in the medium. Hence, the speed of a wave is:

$$v = \frac{\lambda}{T}$$

Combining this equation with the definition of frequency $\left(f = \frac{1}{T}\right)$ yields:

$$v = \frac{\lambda}{T} = f\lambda$$

Therefore, the speed of a wave through a medium can be calculated as either the ratio of *λ* to *T* or the product of *f* and *λ*. It is important to note that this speed only applies to the wave form that is moving through the medium.

Furthermore, every wave moves with the same speed in a medium regardless of its frequency, wavelength, or period. For example, sound waves of various frequencies produced by an orchestra move with the same speed through the air of a concert hall and arrive at a listener's ear simultaneously.

 Concept Check 4.3

A beachball floats on the surface of a lake. A wave moves along the surface of the water and causes the ball to move up and down three times in 12 s. If the distance between the troughs of the wave is 1.6 m, how fast is the wave moving?

Solution

Note: The appendix contains the answer.

4.1.04 Wave Interference

Since **waves** are merely a periodic disturbance in a medium, multiple waves can occupy the same physical space at the same time. When this occurs, the medium takes on the shape that results from adding all the waveforms together. The process of combining waveforms is called **interference**.

When two or more waves occupy the same space in a medium (ie, overlap each other), they combine according to the **principle of superposition** such that the resulting waveform is the sum of each wave's amplitude at every point.

Constructive interference occurs when crests overlap crests (or troughs overlap troughs), and the resultant wave amplitude is increased. Figure 4.8 shows two identical waves adding together to produce a waveform that is double the amplitude of the initial waveforms.

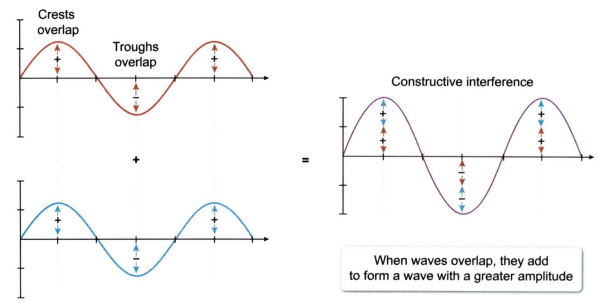

Figure 4.8 Constructive interference occurs when waves combine to produce a larger waveform.

Destructive interference occurs when crests overlap troughs and the resultant wave amplitude is diminished. Parts of a wave above the horizontal axis are considered *positive* whereas parts below the axis are considered *negative*. Figure 4.9 shows two waves combining where the crest from one wave completely cancels the trough from the other. The result is no waveform, and complete destructive interference occurs.

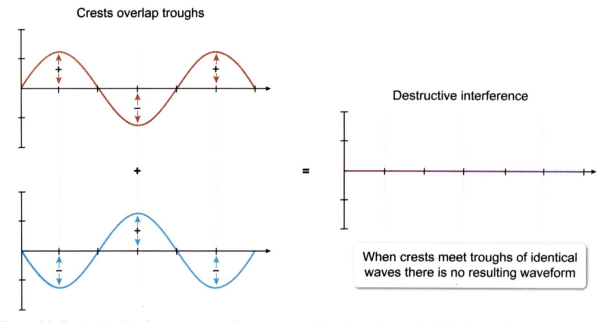

Figure 4.9 Destructive interference occurs when waves combine to produce a diminished waveform.

The shape of the combined waveform is the result of interference that depends on the waves' **phase difference**, which measures the extent to which two waves overlap. If the crests and troughs of one wave line up with the crests and troughs of another wave, the waves are said to be in phase (Figure 4.10). Waves with crests that are separated by $\frac{\lambda}{4}$ are out of phase by 90° and waves with crests that are separated by $\frac{\lambda}{2}$ are out of phase by 180°.

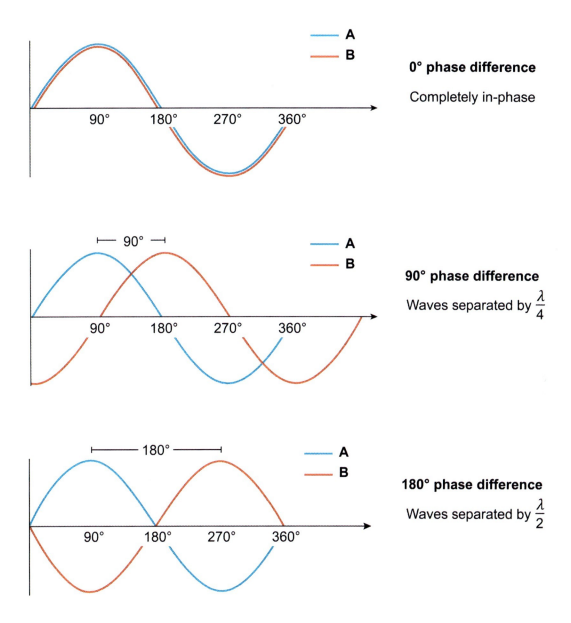

Figure 4.10 Identifying the phase relationship between waves.

When the crests and troughs of two waves do not line up, the waves are partially out of phase. Combine the concepts of interference and phase difference to see the results when waves that are out of phase overlap. For example, where the waves A and B are 90° out of phase (Figure 4.11), some degree of constructive and destructive interference occurs where the two waves overlap to produce the resulting waveform.

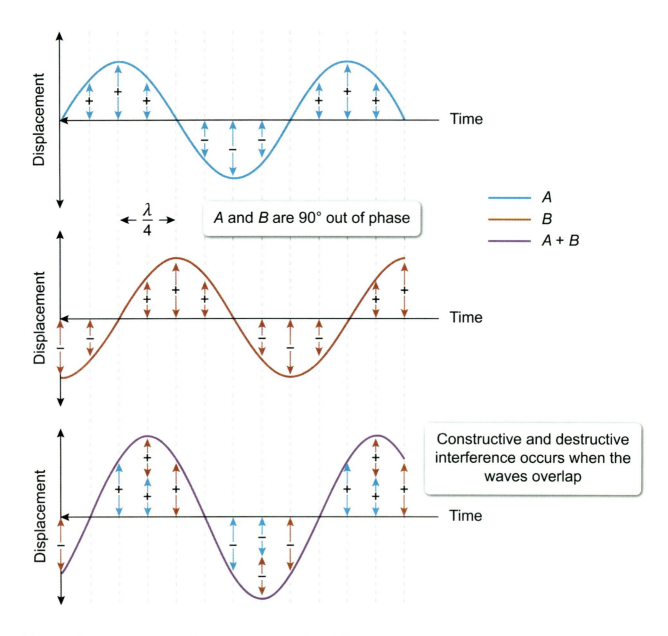

Figure 4.11 Interference formed when two waves that are 90° out of phase overlap.

It is important to note that the process of interference does not alter the original waveforms. The medium takes on the shape of the combined waveforms when two waves overlap. However, after the overlap the original waves continue to move through the medium unaffected by the encounter.

✓ Concept Check 4.4

In the diagram shown, two waves with equal amplitude and frequency overlap. Identify whether constructive or destructive interference occurs at the labeled points *A*, *B*, *C*, *D*, and *E*.

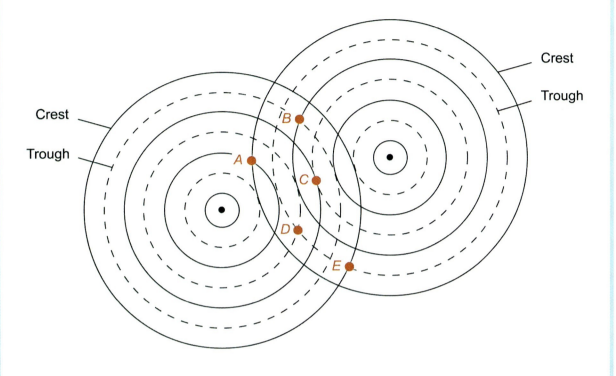

Solution

Note: The appendix contains the answer.

Chapter 4: Waves, Sound, and Light

Lesson 4.2
Sound

Introduction

Sound is a longitudinal wave involving oscillations of pressure in a medium and is commonly encountered in everyday life in the form of speaking, music, thunder, sirens, and similar phenomena. This lesson covers several important physical properties of sound, including its speed, intensity, attenuation, resonance, and the Doppler effect. The lesson concludes with a discussion of biological and medical applications of sound.

4.2.01 Sound Properties

As introduced in Concept 4.1.02, sound is a **mechanical wave** that propagates through a **medium**, such as air, water, and even solids like metals. Figure 4.12 shows sound as a longitudinal wave that displaces the medium in a direction **parallel** to the direction of propagation.

For example, when sound travels through the air, air molecules oscillate back and forth parallel to the direction that the sound travels. Because sound requires oscillations of the medium, sound cannot propagate through a **vacuum**, which is any region devoid of matter including gases, liquids, or solids.

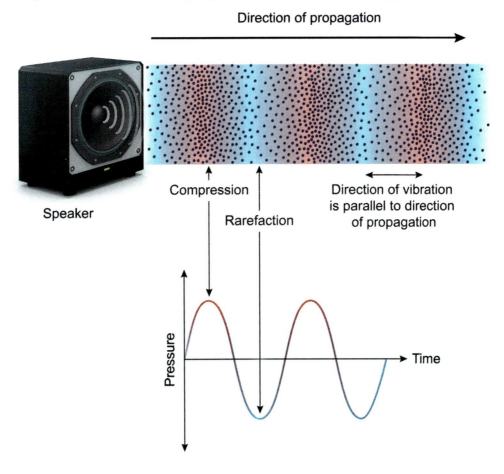

Figure 4.12 Sound is a longitudinal wave that propagates through a medium.

Like other waves discussed in Concept 4.1.03, the speed *v* of sound is equal to the product of its wavelength λ and its frequency *f*:

$$v = \lambda f$$

Sounds with different frequencies are perceived as different **pitches** or **tones** (Figure 4.13). A tuba generates a sound with a low frequency, which is perceived as a low pitch. Alternatively, the singer generates a medium frequency sound (ie medium pitch) and the flute generates a high-frequency sound (ie high pitch). In general, sound waves with frequencies between 20 Hz and 20,000 Hz are within the **audible range** of human hearing.

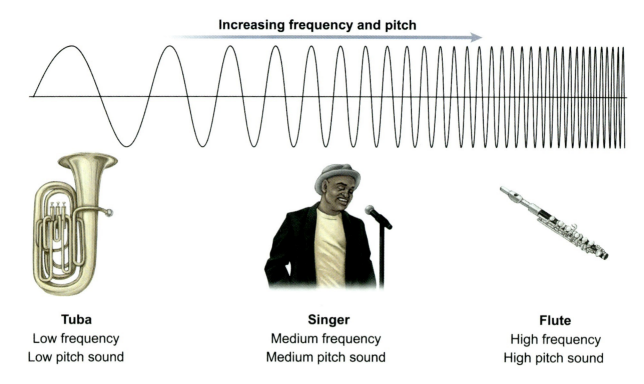

Figure 4.13 The perceived pitch of a sound increases as the frequency increases.

The speed of sound is determined by the physical properties of the medium it travels through, so it is **constant** for a particular medium (Figure 4.14) regardless of the frequency, wavelength, or period of the sound.

Chapter 4: Waves, Sound, and Light

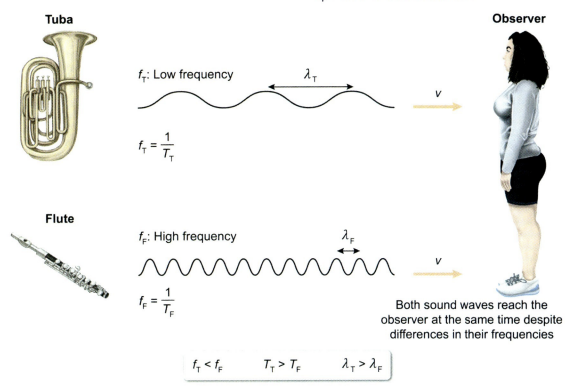

Figure 4.14 The speed of sound is constant in a medium and independent of frequency, wavelength, or period of the sound wave.

One factor that influences the speed of sound is the **temperature** of the medium (see Concept 5.2.01). The speed v of sound in air with a temperature T (in kelvin, K) is calculated as:

$$v = \left(331\ \frac{\text{m}}{\text{s}}\right) \cdot \sqrt{\frac{T}{273\ \text{K}}}$$

Consequently, sound travels *faster* in warmer air and *slower* in cooler air (see Figure 4.15). When the wavelength is constant, an increase (or decrease) in wave speed results in a corresponding increase (or decrease) in the wave's frequency.

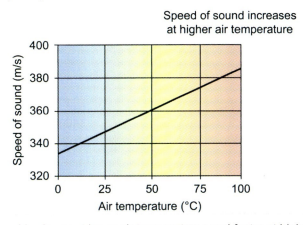

Figure 4.15 The speed of sound is slower at lower air temperatures and faster at higher air temperatures.

In addition, the type of medium also affects the speed of sound. As shown in Figure 4.16, the speed of sound is *slowest* in gases, *faster* in liquids, and *fastest* in solids. Therefore, for a constant wave frequency, the increases in the speed of sound imply that the wavelength is *shortest* in gases, *longer* in liquids, and *longest* in solids.

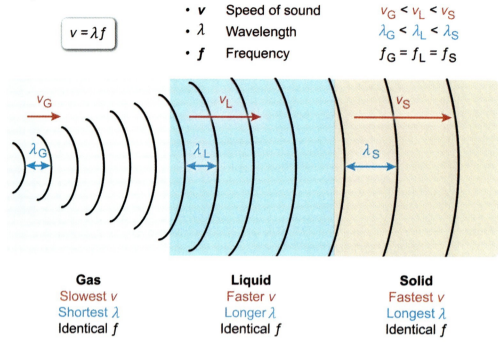

Figure 4.16 The speed and wavelength of sound depends on the medium.

Figure 4.17 demonstrates how to calculate the wavelength of a sound traveling from water into air. In water, the speed of sound equals 1,500 m/s and the frequency equals 500 Hz. In air, the speed of sound equals 340 m/s.

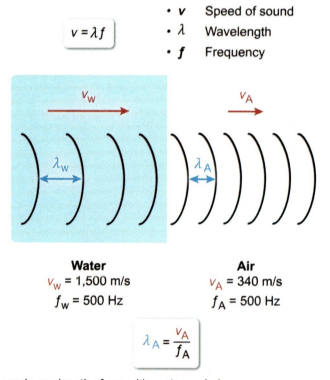

Figure 4.17 The frequency and wavelength of sound in water and air.

Because the frequency of a sound wave does *not depend* on the medium it travels through, the frequency f_A of this sound in air is the same as its frequency f_W in water, 500 Hz:

$$f_A = f_W = 500 \text{ Hz}$$

Furthermore, the speed v_A of sound in air is equal to the product of its wavelength λ_A and f_A:

$$v_A = \lambda_A f_A$$

This equation can be solved for λ_A, yielding:

$$\lambda_A = \frac{v_A}{f_A}$$

Substituting the values for v_A and f_A yields:

$$\lambda_A = \frac{340 \frac{\text{m}}{\text{s}}}{500 \text{ Hz}} \approx 0.7 \text{ m}$$

Therefore, the wavelength of the sound in air is approximately 0.7 m.

Note that when sound encounters an **interface** between two different media, a portion of the sound energy is **reflected**, and the remaining energy is **transmitted** into the new medium. Therefore, sound waves lose energy (are attenuated) each time they pass between different media. Figure 4.18 shows a sound wave traveling through a gas, a liquid, and a solid. Due to reflection at each interface, the energy of the wave transmitted into the new medium is *less than* the energy of the original wave.

As shown in Figure 4.17, when the wave moves from one medium to another, its frequency does not change. Instead, the change in the speed of sound in the new medium changes the wavelength of the sound. In Figure 4.18, the wavelength of the sound increases when it enters the liquid and increases again when it enters the solid. The reflected sound wave has the same v, λ, and f as the original sound wave because it still travels in the **original medium**.

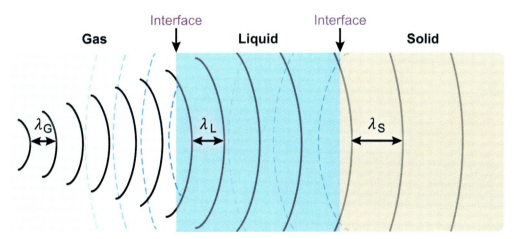

Black sound waves get lighter indicating attenuation due to reflection at the interface
Reflected sound has same speed, λ, and frequency as orginal sound

Figure 4.18 Reflection of sound waves at interfaces of different media cause attenuation.

For example, Figure 4.19 shows sound traveling through the air to the surface of a lake, creating reflected and transmitted sound waves.

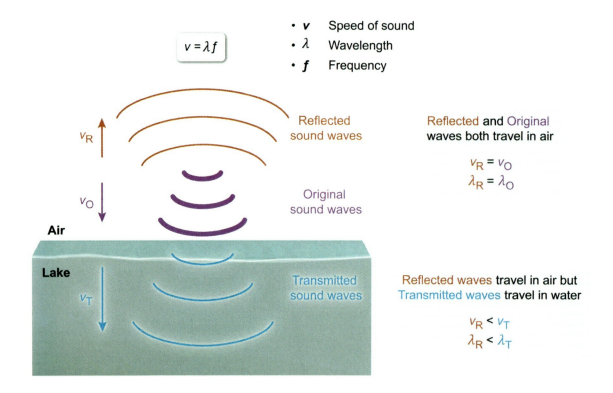

Figure 4.19 Reflected and transmitted sound waves at the surface of a lake.

Because both waves travel in air, the reflected sound wave has the same speed v_R and wavelength λ_R as the original wave's speed v_O and wavelength λ_O:

$$v_R = v_O$$
$$\lambda_R = \lambda_O$$

However, the transmitted wave travels in water instead of air. Therefore, v_R and λ_R are less than the speed v_T and wavelength λ_T of the transmitted wave, respectively:

$$v_R < v_T$$
$$\lambda_R < \lambda_T$$

✓ Concept Check 4.5

A pedestrian hears a car horn with a frequency of 300 Hz and a tornado siren with a frequency of 700 Hz.

   a. Which of these sounds has a higher pitch?
   b. Which of these sounds has a greater speed?

Solution

Note: The appendix contains the answer.

4.2.02 Intensity of Sound

Mechanical waves, such as sound waves, can be described as a **force** that propagates through the medium and displaces molecules within the medium. Consequently, a sound wave does work on the medium and carries **energy** as it propagates.

A sound that causes a larger displacement within the medium does more work, and therefore has more energy (Figure 4.20). As such, the energy E of a mechanical wave is directly proportional to the square of the **amplitude** A of the wave:

$$E \propto A^2$$

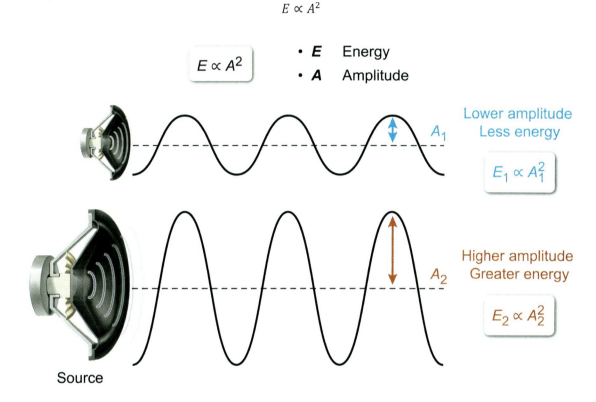

- E Energy
- A Amplitude

Figure 4.20 A sound wave with a higher amplitude has greater energy than a sound wave with a lower amplitude.

The **intensity** I of a sound is defined as the **power** P of the sound wave per unit area, which is equivalent to wave energy E per unit time t per area A (Figure 4.21):

$$I = \frac{P}{A} = \frac{E}{t \cdot A}$$

The SI unit for sound intensity is watts per meter squared (W/m²).

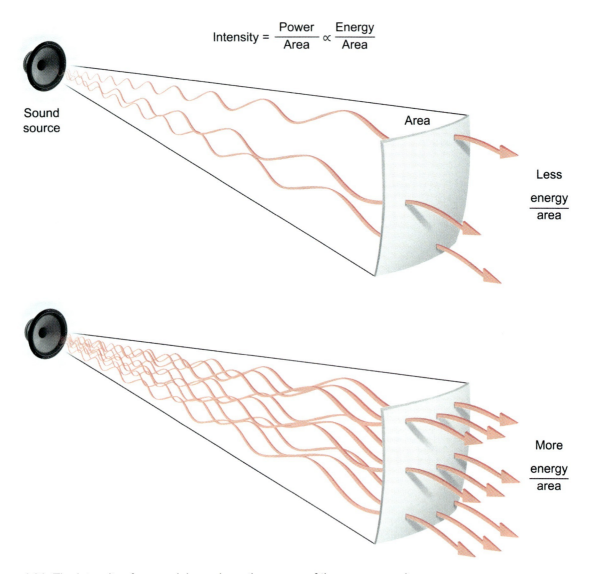

Figure 4.21 The intensity of a sound depends on the energy of the wave per unit area.

The perceived **loudness** of sound is approximately proportional to the logarithm of its intensity, so increasing the intensity of a sound by a factor of 10 is perceived as being twice as loud. Table 4.1 lists useful mathematical properties of logarithms when calculating the loudness of sounds.

Table 4.1 Logarithmic rules for calculating the perceived loudness of sounds in dB.

Product	$\log_{10}(a \cdot b) = \log_{10}(a) + \log_{10}(b)$
Quotient	$\log_{10}\left(\dfrac{a}{b}\right) = \log_{10}(a) - \log_{10}(b)$
Power	$\log_{10}(a^n) = n\log_{10}(a)$
Log of 10	$\log_{10}(10) = 1$
Log of one	$\log_{10}(1) = 0$
Inverse	$\log_{10}\left(\dfrac{1}{a}\right) = -\log_{10}(a)$

The loudness of a sound is measured on the decibel (dB) scale relative to the threshold of normal hearing I_0, which is defined as:

$$I_0 = 1 \times 10^{-12} \, \frac{W}{m^2}$$

The loudness of a sound with intensity I_1 relative to I_0 is calculated in dB with the equation:

$$dB = 10 \log_{10}\left(\frac{I_1}{I_0}\right)$$

Therefore, the loudness of I_1 is 10 dB when it is 10^1 = 10 times more intense than I_0, 20 dB when it is 10^2 = 100 times more intense, 30 dB when it is 10^3 = 1,000 times more intense, and so on.

Furthermore, the difference in loudness in dB between one sound I_1 and another sound I_2 is given by:

$$dB = 10 \log_{10}(I_1) - 10 \log_{10}(I_2)$$

From the quotient rule in Table 4.1, this equation can be expressed as:

$$dB = 10 \log_{10}\left(\frac{I_1}{I_2}\right)$$

For example, the difference in loudness between a 60 dB sound and a 10 dB sound is equal to 50 dB:

$$60 \, dB - 10 \, dB = 50 \, dB$$

Therefore, the 60 dB sound I_1 is 10^5 or 100,000 times more intense than the 10 dB sound I_2:

$$50 \, dB = 10 \log_{10}\left(\frac{I_1}{I_2}\right)$$

$$5 = \log_{10}\left(\frac{I_1}{I_2}\right)$$

$$10^5 = \frac{I_1}{I_2}$$

$$\frac{I_1}{I_2} = 100{,}000$$

Figure 4.22 shows the loudness of thunder and rain are 120 dB and 50 dB, respectively. Based on the definition of the dB scale, the relative intensities of thunder I_T and rain I_R can be calculated as:

$$120 \, dB - 50 \, dB = 10 \log_{10}\left(\frac{I_T}{I_R}\right)$$

$$70 \, dB = 10 \log_{10}\left(\frac{I_T}{I_R}\right)$$

$$7 = \log_{10}\left(\frac{I_T}{I_R}\right)$$

$$10^7 = \frac{I_T}{I_R}$$

$$\frac{I_T}{I_R} = 10{,}000{,}000$$

Therefore, the sound intensity of thunder is 10 million times greater than the sound intensity of rain, which corresponds to a 70 dB difference in the loudness of these two sounds.

$$dB = 10 \log_{10}\left(\frac{I_1}{I_2}\right)$$

- I_1 Intensity of sound 1
- I_2 Intensity of sound 2

Thunder (I_1): 120 dB   **Rain (I_2): 50 dB**

120 dB − 50 dB = 70 dB

$$70\ dB = 10 \log_{10}\left(\frac{I_1}{I_2}\right)$$

$$\frac{I_1}{I_2} = 10^7 = 10{,}000{,}000$$

Sound of thunder is 10 million times more intense than sound of rain

Figure 4.22 The sound intensity of thunder is 10 million times greater than the sound intensity of rain.

As shown in Figure 4.21, the intensity *I* of a sound wave is equal to the power *P* per unit area *A*:

$$I = \frac{P}{A}$$

As a sound wave propagates away from the source in all directions, the power of the wave spreads out along an ever-increasing area. Although the total **power** of the wave stays **constant**, the intensity of the sound—and therefore the perceived loudness—decreases as the **distance** from the source increases. The increasing area occupied by the sound wave represents a sphere with an increasing radius *r* relative to the source and a surface area *A* given by:

$$A = 4\pi r^2$$

Substituting this relationship into the equation for intensity shows *I* and *r* are related by:

$$I = \frac{P}{4\pi r^2} = \left(\frac{P}{4\pi}\right)\left(\frac{1}{r^2}\right)$$

The value of $\frac{P}{4\pi}$ remains constant as the sound wave propagates. However, the intensity *I* decreases according to the square of *r*:

$$I \propto \frac{1}{r^2}$$

Therefore, the intensity of a sound projected in all directions is inversely proportional to the square of the distance from the sound's source (Figure 4.23).

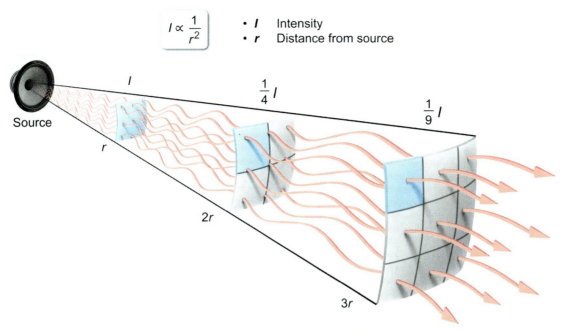

Figure 4.23 The intensity of a sound decreases relative to the square of the distance from the source.

For example, Figure 4.24 shows a jet engine emitting a sound of 150 dB when an airplane flies at an altitude of 10,000 m. Because of the distance r to the ground, the sound intensity I_G observed at ground level is less than the sound intensity I_A emitted by the engine by a factor of 10^8:

$$I_G = \frac{I_A}{r^2} = \frac{I_A}{(10,000)^2} = \frac{I_A}{10^8}$$

- I Intensity
- r Distance from source

$I \propto \frac{1}{r^2}$

Airplane (I_A): 150 dB

Altitude: $r = 10,000$ m

Ground (I_G)

Change in sound intensity of the engine at ground level

$$I_G = \frac{I_A}{r^2} = \frac{I_A}{(10,000)^2}$$

Change in loudness of the engine at ground level

$$10 \log_{10}\left(\frac{I_G}{I_A}\right) = 10 \log_{10}\left(\frac{\frac{I_A}{10^8}}{I_A}\right)$$

Figure 4.24 Loudness at ground level of a jet engine in flight.

The decrease in the loudness of the engine at ground level in dB is calculated as:

$$10\log_{10}\left(\frac{I_G}{I_A}\right) = 10\log_{10}\left(\frac{\frac{I_A}{10^8}}{I_A}\right) = 10\log_{10} 10^{-8}$$

$$10\log_{10} 10^{-8} = 10(-8)\text{ dB} = -80\text{ dB}$$

When the sound of the engine reaches ground level, the loudness equals the sum of the loudness at the airplane and the change in loudness:

$$150\text{ dB} + (-80\text{ dB}) = 70\text{ dB}$$

Therefore, the loudness of the jet engine's sound when it reaches ground level is 70 dB.

> **Concept Check 4.6**
>
> A sound is 100,000 times more intense than the limit of human hearing. How loud is the sound in dB?
>
> **Solution**
>
> *Note: The appendix contains the answer.*

4.2.03 Resonance

When a wave is restricted to a region with **boundaries**, such as a string (like a violin string) fixed at both ends, the incident and reflected waves can perfectly overlap each other and form a standing wave (Figure 4.25). This wave pattern results from the combined effects of constructive and destructive interference described in Concept 4.1.04.

Standing waves appear to stay in place and have **nodes** and **antinodes** that are stationary. Nodes are locations where the standing wave has **zero** displacement (zero wave amplitude), and antinodes are locations with **maximum** displacement (maximum wave amplitude).

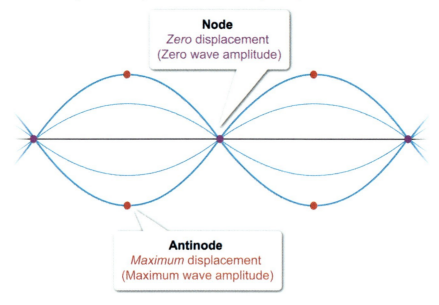

Figure 4.25 A standing wave consists of stationary points with zero displacement (nodes) and maximum displacement (antinodes).

Standing waves only form at a certain set of **harmonic** frequencies, where constructive interference increases the wave **amplitude** and causes resonance. The specific resonant frequencies depend on the physical characteristics and dimensions of the object.

The **fundamental** frequency is the lowest possible resonant frequency for a particular object, denoted with a **harmonic number** n equal to 1. Subsequent harmonics occur at integer **multiples** of the fundamental frequency (usually n = 2, 3, 4…). Similarly, the fundamental wavelength is the longest possible resonant wavelength for an object and is denoted as n = 1, with subsequent harmonics occurring at integer **divisions** of the fundamental wavelength (typically n = 2, 3, 4…).

The exam covers standing waves on three specific structures:

- a **string** fixed at both ends
- sound produced by air in a **pipe** with both ends open
- sound produced by air in a pipe with one end open and one end closed

For each possible harmonic, the speed v of the waves is equal to the product of the harmonic wavelength λ_n and the harmonic frequency f_n:

$$v = \lambda_n f_n$$

Standing waves can form on a string fixed at both ends at specific harmonic frequencies (n = 1, 2, 3…). The frequency and wavelength for each harmonic are related to the dimensions of the string (Figure 4.26).

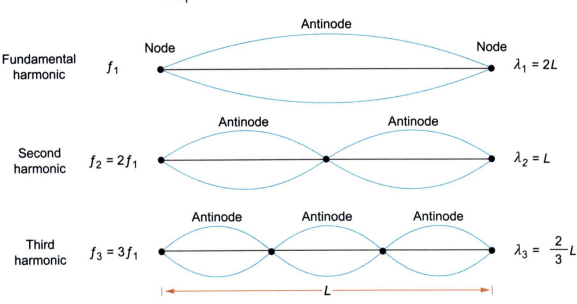

Figure 4.26 Frequencies and wavelengths of the first three harmonics on a string.

The fundamental harmonic (n = 1) of a string has one antinode and a wavelength λ_1 equal to twice the length L of the string:

$$\lambda_1 = 2L$$

The fundamental frequency f_1 is equal to the ratio of the wave speed v to λ_1:

$$f_1 = \frac{v}{\lambda_1} = \frac{v}{2L}$$

The wavelength λ_n of the nth harmonic is equal to integer divisions of λ_1:

$$\lambda_n = \frac{2L}{n}$$

where n equals any positive integer.

Alternatively, the frequency f_n of the nth harmonic is equal to integer multiples of f_1:

$$f_n = nf_1 = \frac{nv}{2L}$$

Note that as n increases, λ_n decreases and f_n increases. Furthermore, the number of antinodes on the string corresponds to the value of n.

For example, Figure 4.27 shows a 400 Hz standing wave on a 60 cm guitar string. The diagram of the standing wave shows two antinodes, corresponding to a harmonic number $n = 2$.

$\lambda_n = \frac{2}{n}L$

$f_n = \frac{nv}{\lambda_1}$

$v = \frac{\lambda_1 f_n}{n}$

- λ Wavelength
- L Length of string
- n Integer multiple (n = 1, 2, 3,...)
- f Frequency
- v Wave speed

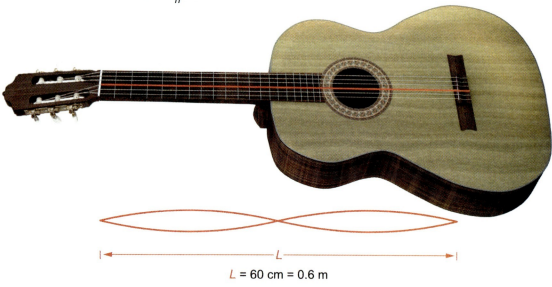

L = 60 cm = 0.6 m

Two antinodes on standing wave: $n = 2$ $\lambda_1 = \frac{2}{1}L$

Standing wave frequency is 400 Hz: $f_2 = 400$ Hz $v = \frac{\lambda_1 f_2}{n}$

Figure 4.27 Standing wave on a guitar string with length 60 cm.

Hence, the frequency of the standing wave equals the second harmonic frequency f_2:

$$f_2 = 400 \text{ Hz}$$

Furthermore, the wavelength λ_1 of the fundamental harmonic equals twice the length of the string L:

$$\lambda_1 = 2L = 2(0.6 \text{ m}) = 1.2 \text{ m}$$

The nth harmonic frequency f_n is calculated from the harmonic number n, the fundamental wavelength λ_1, and the wave speed v:

$$f_n = \frac{nv}{\lambda_1}$$

Solving this equation for v yields:

$$v = \frac{\lambda_1 f_n}{n}$$

Substituting the values for each variable gives:

$$v = \frac{(1.2 \text{ m})\left(400\,\frac{1}{\text{s}}\right)}{2} = 240\,\frac{\text{m}}{\text{s}}$$

Therefore, the wave speed on the guitar string is 240 m/s.

In addition, the speed of a wave on a string is equal to the square root of the ratio of the string **tension** force F_T and the linear **density** μ of the string (ie, its mass per unit length):

$$v = \sqrt{\frac{F_T}{\mu}}$$

Therefore, the harmonic frequencies are directly proportional to the square root of the tension force and inversely proportional to the square root of the linear density:

$$f_n = \frac{1}{\lambda_n}\sqrt{\frac{F_T}{\mu}}$$

 Concept Check 4.7

To tune a guitar, a musician checks the resonant frequency of each string. One string produces a lower frequency than needed, so the musician turns the tuning peg for that string. How could turning the tuning peg increase the resonant frequency of the guitar string?

Turning a tuning peg on a guitar

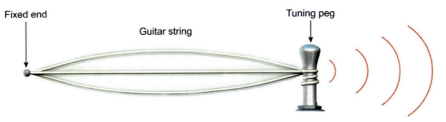

Resonant frequency is too low

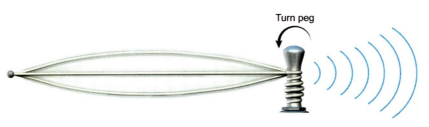

Resonant frequency increases

Solution
Note: The appendix contains the answer.

Standing waves can also be formed in the air inside **pipes** at specific harmonic frequencies where the incident and reflected waves perfectly overlap. The frequency and wavelength of each harmonic are related to the dimensions of the pipe but also depend on whether the ends of the pipe are open or closed. The **open end** of a pipe is the location of an **antinode** (maximum displacement) in the standing wave and the **closed end** is the location of a **node** (zero displacement), as shown in Figure 4.28.

For a pipe open at **both ends**, the wavelength λ_1 of the fundamental harmonic is equal to twice the length L of the pipe:

$$\lambda_1 = 2L$$

The fundamental frequency f_1 is equal to the ratio of the wave speed v to λ_1:

$$f_1 = \frac{v}{\lambda_1} = \frac{v}{2L}$$

The wavelength λ_n and frequency f_n of the nth harmonic (n = 1, 2, 3…) are calculated from the equations:

$$\lambda_n = \frac{2L}{n}$$

$$f_n = nf_1 = \frac{nv}{2L}$$

where n equals any positive integer. Note that as n increases, λ_n decreases and f_n increases. Furthermore, the number of nodes in the pipe corresponds to the value of n.

Alternatively, a pipe open at **one end** and closed at the other end has a fundamental harmonic with a wavelength λ_1 equal to four times the length L of the pipe:

$$\lambda_1 = 4L$$

The fundamental frequency f_1 is equal to the ratio of the wave speed v to λ_1:

$$f_1 = \frac{v}{\lambda_1} = \frac{v}{4L}$$

The wavelength λ_n and frequency f_n of the nth harmonic are calculated from the equations:

$$\lambda_n = \frac{4L}{n}$$

$$f_n = nf_1 = \frac{nv}{4L}$$

where the value of n equals only *odd* positive integers (n = 1, 3, 5…). Note that as n increases, λ_n decreases and f_n increases. However, the value of n *does not* correspond to the number of nodes or antinodes in the pipe.

Open on both ends

$\lambda_1 = 2L \quad \lambda_n = \dfrac{2L}{n} \quad f_n = \dfrac{nv}{2L}$

- λ Wavelength
- L Length of pipe
- n Integer multiple ($n = 1, 2, 3,...$)

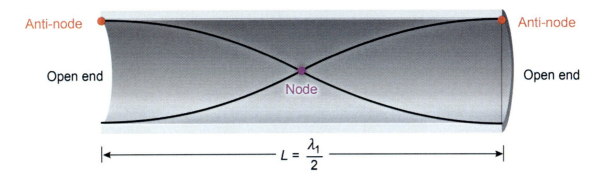

Open on one end

$\lambda_1 = 4L \quad \lambda_n = \dfrac{4L}{n} \quad f_n = \dfrac{nv}{4L}$

- λ Wavelength
- L Length of pipe
- n Odd integers ($n = 1, 3, 5,...$)

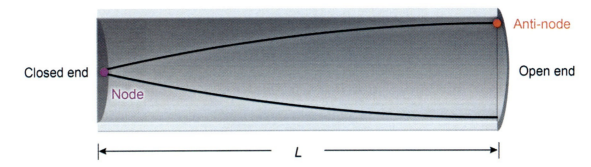

Figure 4.28 Standing waves in a pipe open on both ends compared to a pipe open on only one end.

Another physical property of a pipe that affects the frequency of the standing waves is the wave speed v for the medium in the pipe. The speed of a wave in a pipe is inversely proportional to the square root of the **density** ρ of the gas within the pipe:

$$v \propto \dfrac{1}{\sqrt{\rho}}$$

Because the harmonic frequencies f_n are proportional to v, they are also inversely proportional to the square root of ρ:

$$f_n \propto v \propto \dfrac{1}{\sqrt{\rho}}$$

For example, the speed of sound in air is 330 m/s whereas the speed of sound in helium gas, which is less dense than air, is equal to 960 m/s. Standing waves at the fundamental frequencies are generated in two 0.3 m pipes that are open on both ends. One of the pipes is filled with air and the other pipe is filled with helium (Figure 4.29).

$\lambda_1 = 2L$

$f_1 = \dfrac{v}{2L}$

- λ_1 Fundamental wavelength
- L Length of pipe
- f_1 Fundamental frequency
- v Wave speed

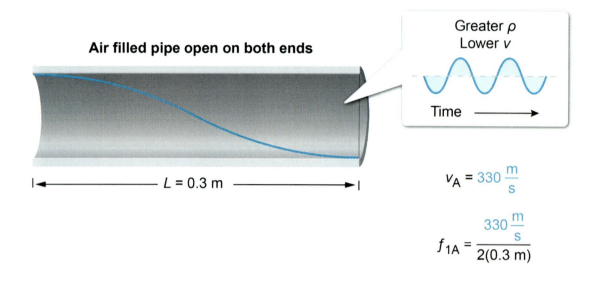

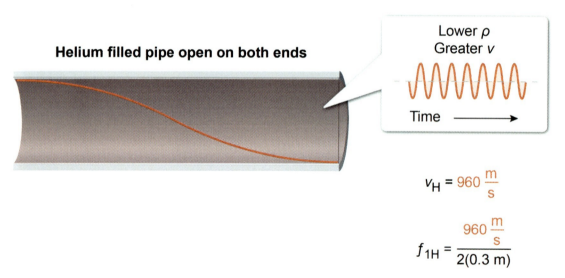

Figure 4.29 The fundamental frequency of a pipe depends on the density of the gas in the pipe due to differences in wave speed.

A pipe open on both ends and with a length L of 0.3 m has a fundamental harmonic with a wavelength λ_1 given by:

$$\lambda_1 = 2L = (2)(0.3 \text{ m}) = 0.6 \text{ m}$$

If the pipe is filled with air, the fundamental frequency f_{1A} equals the ratio of the wave speed v_A in air to λ_1:

$$f_{1A} = \dfrac{v_A}{\lambda_1} = \dfrac{330 \, \frac{\text{m}}{\text{s}}}{0.6 \text{ m}} = 550 \, \dfrac{1}{\text{s}} = 550 \text{ Hz}$$

However, if the pipe is filled with helium, the fundamental frequency f_{1H} increases due to the increased wave speed v_H in helium:

$$f_{1H} = \frac{v_H}{\lambda_1} = \frac{960 \frac{m}{s}}{0.6 \, m} = 1{,}600 \, \frac{1}{s} = 1{,}600 \, \text{Hz}$$

 Concept Check 4.8

A pipe of length 0.8 m is open at both ends and produces a standing wave with a wavelength of 0.2 m. How many nodes and antinodes does the standing wave have?

Solution

Note: The appendix contains the answer.

4.2.04 Doppler Effect

The Doppler effect causes the frequency of a sound emitted from a **source** to be perceived as a different frequency by an **observer** when the source and/or observer are moving relative to each other (Figure 4.30). This change in frequency occurs because the time between adjacent sound wave fronts increases or decreases depending on the relative motion of the source and observer.

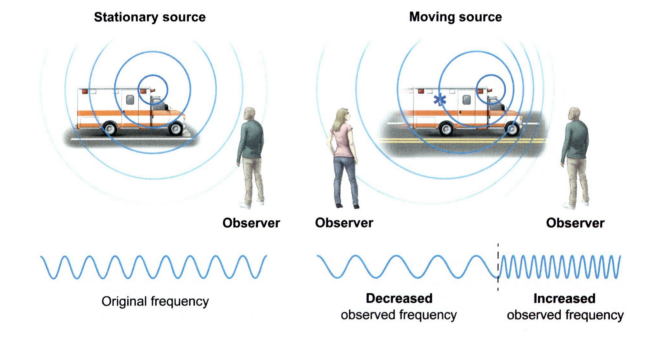

Figure 4.30 The Doppler effect changes the observed frequency of a sound due to the motion of the source and/or the observer.

Chapter 4: Waves, Sound, and Light

When the source and observer move **toward** each other, the perceived frequency is *greater* than the source frequency. However, when the source and observer move **away** from each other, the perceived frequency is *less* than the source frequency. The perceived sound frequency *f* depends on the speed of sound *v*, the observer velocity v_o, the source velocity v_s, and the frequency f_s of the sound transmitted by the source according to the formula:

$$f = \left(\frac{v \pm v_o}{v \mp v_s}\right) f_s$$

Table 4.2 shows that a **plus sign** (top sign) is used in the **numerator** when the observer is moving *toward* the source, and a **minus sign** (bottom sign) is used when the observer is moving *away* from the source. Similarly, the top sign (minus) is used in the **denominator** when the source is moving *toward* the observer, and the bottom sign (plus) is used when the source is moving *away* from the observer.

Table 4.2 Doppler effect equations for relative motion between the source and observer.

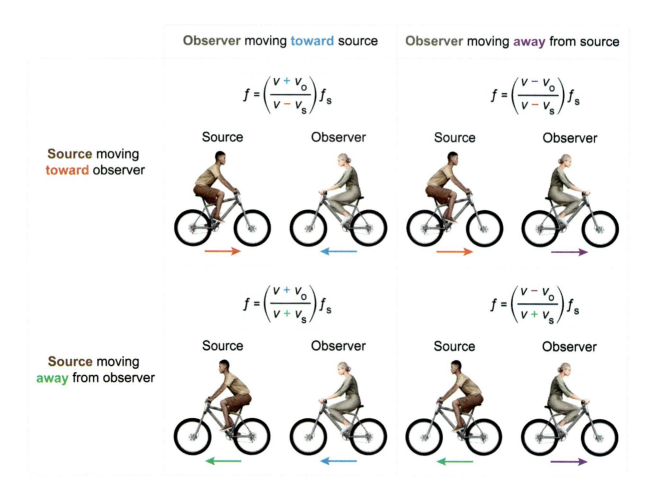

Note that only the direction of the movement between the source or the observer is relevant, not whether the distance between them is increasing or decreasing. Furthermore, the Doppler effect *does not* change the speed of sound and *does not* depend on the distance between the source and the observer.

For example, two cars travel eastward at 20 m/s, as shown in Figure 4.31. The first car sounds its horn with a frequency of 500 Hz and with the speed of sound equal to 330 m/s.

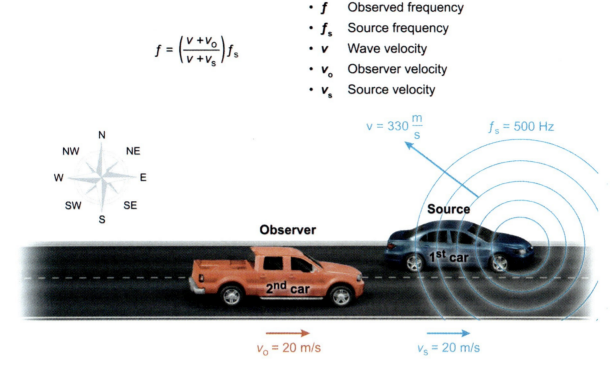

Figure 4.31 No Doppler effect is perceived when no relative motion exists between the source and the observer.

The first car is the source of the sound and travels away from the second car, which is the observer. Furthermore, the observer travels toward the source. As a result, a plus sign is used in the numerator of the Doppler effect equation, and a plus sign is used in the denominator.

The frequency f of the horn perceived by the observer is calculated from the speed of sound v (330 m/s), the observer velocity v_o (20 m/s), the source velocity v_s (20 m/s), and the frequency f_s of the horn (500 Hz) at the source:

$$f = \left(\frac{v + v_o}{v + v_s}\right) f_s = \left(\frac{330 \frac{m}{s} + 20 \frac{m}{s}}{330 \frac{m}{s} + 20 \frac{m}{s}}\right) 500 \text{ Hz}$$

$$f = \left(\frac{350 \frac{m}{s}}{350 \frac{m}{s}}\right) 500 \text{ Hz} = 500 \text{ Hz}$$

Therefore, the observed frequency of the horn is 500 Hz, and no Doppler effect is perceived in this situation because no relative motion exists between the two cars.

As another example, consider a car traveling at 30 m/s toward a truck that is traveling 50 m/s away from the car (Figure 4.32). The car sounds its horn with a frequency of 800 Hz, and the speed of sound is equal to 330 m/s.

$$f = \left(\frac{v - v_o}{v - v_s}\right) f_s$$

- f Observed frequency
- f_s Source frequency
- v Wave velocity
- v_o Observer velocity
- v_s Source velocity

$f_s = 800$ Hz $v = 330 \frac{m}{s}$

Source   Observer

$v_s = 30$ m/s $v_o = 50$ m/s

Truck (observer) moving away from car: $v_o = 50$ m/s
Car (source) moving toward truck: $v_s = 30$ m/s

$$f = \left(\frac{v - 50 \frac{m}{s}}{v - 30 \frac{m}{s}}\right)(800 \text{ Hz})$$

Figure 4.32 Doppler effect for a car sounding its horn with a frequency of 800 Hz.

The truck is the observer and has a velocity v_o of 50 m/s. A minus sign is used in the numerator because the truck is moving away from the source. The car is the source and is moving toward the observer, so a minus sign is also used in the denominator. Substituting the values and proper signs into the Doppler effect equation yields:

$$f = \left(\frac{v - v_o}{v - v_s}\right) f_s = \left(\frac{330 \frac{m}{s} - 50 \frac{m}{s}}{330 \frac{m}{s} - 30 \frac{m}{s}}\right) 800 \text{ Hz}$$

$$f = \left(\frac{280 \frac{m}{s}}{300 \frac{m}{s}}\right) 800 \text{ Hz} \approx (0.93) 800 \text{ Hz} \approx 740 \text{ Hz}$$

Therefore, the truck observes the frequency of the car horn to be 740 Hz because the truck's relative motion away from the car causes the perceived frequency to be less than the source frequency.

 Concept Check 4.9

An ambulance siren has a frequency of 1,000 Hz. If a stationary pedestrian perceives the siren to have a frequency of 1,250 Hz, is the ambulance moving toward the pedestrian or away from them?

Solution

Note: The appendix contains the answer.

4.2.05 Applications of Sound

The properties of sound waves—including formation, interference, reflection, resonance, and the Doppler effect—play significant roles in many biological processes and medical procedures. Sound waves are especially important in human speech, ultrasound imaging, and extracorporeal shock wave lithotripsy.

Speech

The sound produced by human vocal cords is similar to the standing waves produced on a string fixed on both ends (see Concept 4.2.03). Changing the length, tension, or linear density of the vocal cords changes the frequency of the sound produced. Males generally have vocal cords that are longer and thicker than females' vocal cords, resulting in male voices typically having a lower frequency and lower pitch than female voices.

Ultrasound Imaging

Ultrasound imaging is used to map structures within the body using sound frequencies much higher than the range of human hearing, up to 15 MHz. An ultrasound **probe** transmits sound pulses into the body and records the sound waves that reflect off different anatomic structures.

As discussed in Concept 4.2.01, when sound encounters an interface between two different media, a portion of the sound energy is reflected and forms an **echo**. The **interfaces** between different tissues (eg, fat, muscle, blood, organs) generate echoes that reflect back to the imaging probe. The ultrasound system uses this echo data to reconstruct an image of different tissues within the body (see Figure 4.33).

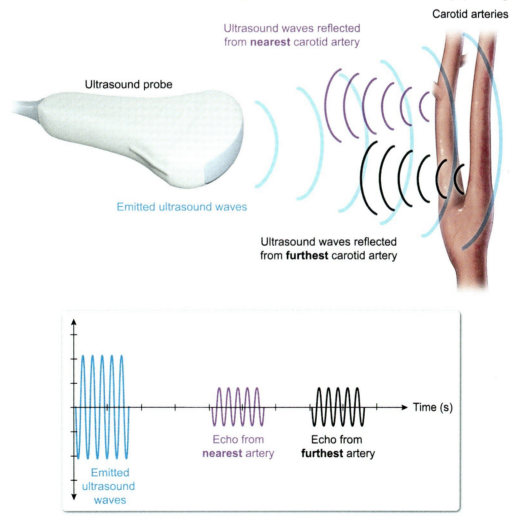

Figure 4.33 Ultrasound imaging records echo signals from interfaces between different tissues to map structures inside the body.

The **elapsed time** that the wave travels through the body includes both the time to reach the tissue and the time for the reflected sound to return back to the probe. The distance *d* from the ultrasound probe to a particular structure is equal to half the product of the speed of sound *v* in the body tissue and the elapsed time Δ*t*:

$$d = \frac{v \cdot \Delta t}{2}$$

Concept 4.2.04 describes how relative motion between a sound source and an observer leads to a Doppler shift in the frequency of the sound. This principle allows ultrasound imaging to measure the speed of **blood flow** within the body (Figure 4.34). Doppler ultrasound measures the Doppler shift Δ*f* of the sound waves, which is equal to the difference between the observed frequency *f* of the echoes and the source frequency f_s:

$$\Delta f = f - f_s$$

Because the blood velocity v_{blood} is much less than the speed of sound *v*, Δ*f* can be approximated as:

$$\Delta f \approx \frac{v_{blood} \cdot f_s}{v}$$

This equation can be rearranged to show that the ratio of Δ*f* and f_s is approximately equal to the ratio of v_{blood} and *v*:

$$\frac{\Delta f}{f_s} \approx \frac{v_{blood}}{v}$$

Similarly, the ratio of the change in wavelength Δ*λ* and the source wavelength λ_s is approximately equal to the ratio of v_{blood} and *v*:

$$\frac{\Delta \lambda}{\lambda_s} \approx \frac{v_{blood}}{v}$$

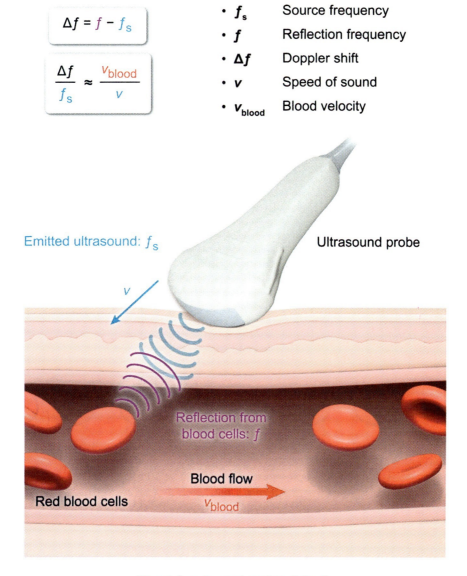

Figure 4.34 Doppler ultrasound measures blood velocity from the frequency shift of reflected ultrasound waves.

However, Doppler ultrasound uses the probe to both transmit the sound wave as well as to detect Δf. As a result, the sound wave frequency is shifted twice. The frequency shifts once when the ultrasound waves encounter the flowing blood (the ultrasound probe acts as the source, and the flowing blood acts as the observer). The frequency then shifts a second time when the ultrasound waves reflect back to the probe (the flowing blood acts as the source, and the ultrasound probe acts as the observer).

Shock Wave Lithotripsy

Ultrasound waves can also be used during therapeutic interventions, such as extracorporeal shock wave lithotripsy (Figure 4.35). This treatment uses focused ultrasound to break kidney stones into small fragments that can be cleared from the body through the urine. These ultrasound waves converge at a **focal point** within the body where the therapeutic effect is desired. The focal point experiences abrupt pressure changes due to the formation of shock waves.

Chapter 4: Waves, Sound, and Light

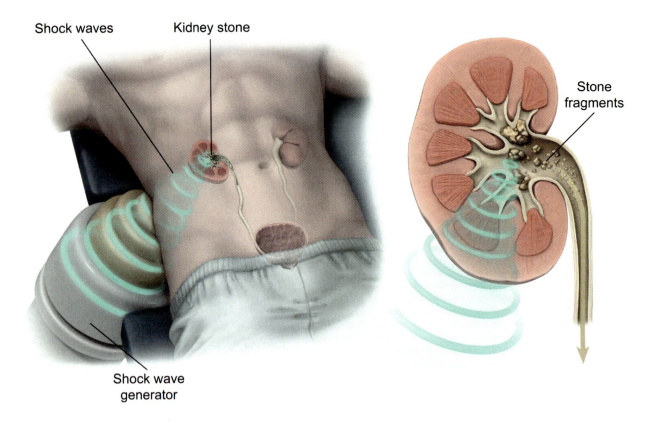

Figure 4.35 Focused ultrasound can pulverize kidney stones during extracorporeal shock wave lithotripsy.

Shock waves are extremely brief (approximately discontinuous) changes in the pressure and density within a medium. The most familiar examples of shock waves occur when objects move faster than the speed of sound in the medium. For example, shock waves are associated with the sonic boom heard as a supersonic aircraft passes above or the crack of a whip when its end moves faster than the speed of sound in air. In the kidney stone, the abrupt changes in pressure break the stone into pieces.

In addition, the frequency of the ultrasound is adjusted to match the **resonant frequency** of the kidney stones. Like the strings and pipes in Concept 4.2.03, standing waves can be generated in the solid structure of a kidney stone, causing constructive wave interference. This greatly increases the wave **amplitude** and therefore the **energy** deposited into the stone, which breaks the stone into smaller fragments.

The ultrasound waves do not resonate in the kidney or surrounding tissues because of the differences in their structure (liquid versus solid) and the relative speed of sound.

 Concept Check 4.10

An ultrasound probe records an echo from a tissue interface 0.16 ms after emitting a pulse. If the tissue interface is 10 cm from the probe, what is the speed of sound?

Solution

Note: The appendix contains the answer.

Lesson 4.3
Light

Introduction

This lesson focuses on the topic of the electromagnetic radiation known as light. First, the properties of light waves and the electromagnetic spectrum (ie, the range of all types of electromagnetic radiation) are covered. Next, the behavior of light waves is discussed, including their properties and the various phenomena that can occur when light interacts with matter: reflection, refraction, polarization, and dispersion. This lesson also covers the photoelectric effect, which describes the emission of electrons from a material when it is exposed to light. Lastly, the Doppler effect (as it pertains to light) is introduced.

4.3.01 Light Waves

The field of optics, which covers the study of light and its behavior, plays a significant role in several areas of medicine. As such, understanding the behavior of **light waves** is crucial in understanding the properties of lenses, mirrors, and other objects through which light interacts. For example, lenses can focus light waves to create images, and mirrors can reflect and redirect light waves.

These waves of light are a form of **electromagnetic radiation**, which refers to the propagation of waves consisting of oscillating electric and magnetic fields (Figure 4.34). In electromagnetic waves, the direction of the electric field (E) and the magnetic field (B) vectors are perpendicular to one another and perpendicular to the direction of wave propagation.

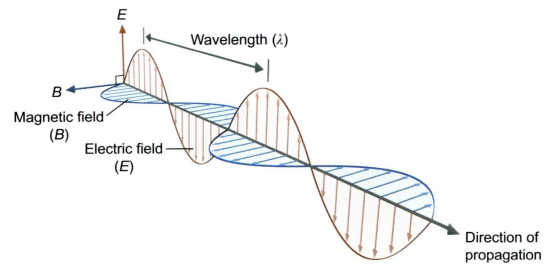

Figure 4.34 Light waves consist of electric and magnetic fields oscillating perpendicular to one another.

The **electromagnetic spectrum** refers to all classifications of radiation, which differ from one another based on the **frequency** f (in hertz, Hz) and **wavelength** λ (in meters, m) of the oscillating electric and magnetic fields. However, all light waves are related in the respect that they travel at the same speed c (ie, the **speed of light**) in a vacuum, equal to **3 × 10⁸ m/s**. Moreover, this speed is always equal to the product of λ and f:

$$c = \lambda f = 3 \times 10^8 \text{ m/s}$$

Consequently, as the wavelength of the light wave increases, the frequency must decrease for the product of the two to remain constant and equal to c.

> **Concept Check 4.11**
>
> A beam of blue light is measured to have a wavelength of 400 nm in a vacuum. What is the frequency of the blue light wave?
>
> **Solution**
>
> *Note: The appendix contains the answer.*

4.3.02 Electromagnetic Spectrum

The **electromagnetic spectrum** describes classifications of electromagnetic radiation that differ from one another in terms of waveform characteristics, occurrence in the natural world, and technological utility. In the context of the exam, the focus is on the regions of the electromagnetic spectrum that are most relevant to medical applications.

Visible light is the region of the electromagnetic spectrum that is most relevant to human vision. It has a wavelength range of approximately 400 to 700 nanometers (nm) and is responsible for the colors we see in the world around us. The color red has the longest wavelength (ie, lowest energy), and the color blue has the shortest wavelength (ie, highest energy), as shown in Figure 4.35.

In addition to its role in vision, visible light is also important in a variety of medical applications. For example, it is used in optical imaging techniques such as microscopy and endoscopy, as well as in radiation therapy.

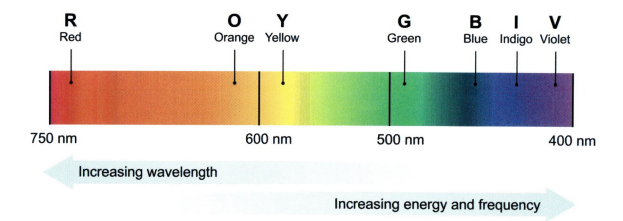

Figure 4.35 The visible portion of the electromagnetic spectrum includes the colors detectable by the human eye.

Ultraviolet (UV) radiation is the region of the electromagnetic spectrum with shorter wavelengths than visible light. UV wavelengths are in the range of 10 nm to 400 nm and are divided into three subregions: UV-A, UV-B, and UV-C. Although UV radiation can cause damage to biological tissues, it also has medical applications, such as in phototherapy for skin conditions like psoriasis.

Infrared (IR) radiation has longer wavelengths than visible light, in the range of 700 nm to 1 mm, and is divided into three subregions: near-infrared, mid-infrared, and far-infrared. IR radiation is used in medical imaging techniques, such as thermal imaging and infrared spectroscopy, as well as laser surgery and physiotherapy.

X-rays and **gamma rays** include the regions of the electromagnetic spectrum with the shortest wavelengths and highest frequencies. X-rays have wavelengths in the range of 0.01 nm to 10 nm, and gamma rays have wavelengths less than 0.01 nm. Both X-rays and gamma rays are ionizing radiation, meaning they can cause damage to biological tissues by stripping electrons from atoms. However, they also have important medical applications, such as in diagnostic imaging and radiation therapy.

The electromagnetic spectrum is typically arranged in order of frequency f or wavelength λ. Because the frequency and wavelength of electromagnetic radiation are inversely proportional, arranging the electromagnetic spectrum from longest wavelength (radio waves) to shortest wavelength (gamma rays) is equivalent to arranging the electromagnetic spectrum from smallest frequency to largest frequency (Figure 4.36).

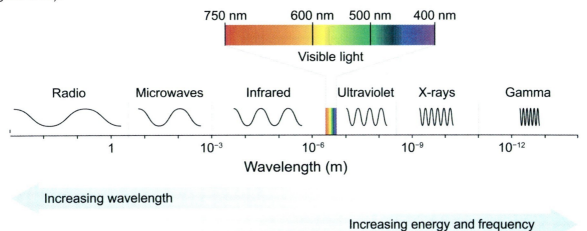

Figure 4.36 The electromagnetic spectrum is categorized into several regions and ordered by wavelength or frequency.

Alternatively, the electromagnetic spectrum may also be ordered in terms of the **energy** E associated with a given electromagnetic oscillation, which can best be described in terms of the particle (ie, photon) nature of light. The photon is a massless, neutral particle that mediates electromagnetic interactions, and can be thought of as representing a wave packet of electromagnetic energy.

This energy is described by the Planck-Einstein equation, which relates the energy E of an electromagnetic wave (and corresponding photon) to wave frequency and **Planck's constant** ($h = 6.62 \times 10^{-34}$ m²·kg/s):

$$E = hf$$

By using the equation for wave speed, where the speed of light $c = \lambda f$, and solving for the frequency, the equation can be written as:

$$E = \frac{hc}{\lambda}$$

Accordingly, the energy of electromagnetic radiation is *inversely proportional* to wavelength and *directly proportional* to frequency. Thus, the order of electromagnetic radiation types arranged from lowest energy to highest energy is:

$$E_{\text{radio waves}} < E_{\text{microwaves}} < E_{\text{infrared light}} < E_{\text{visible light}} < E_{\text{ultraviolet light}} < E_{\text{x-rays}} < E_{\text{gamma rays}}$$

The **intensity** I of electromagnetic waves is defined as the power P (or energy E per unit time t) delivered per unit area A:

$$I = \frac{P}{A} = \frac{E}{t \cdot A}$$

In other words, the intensity of electromagnetic radiation (including light) depends on the energy of each individual photon and the number of photons per unit time.

> ☑ **Concept Check 4.12**
>
> What is the energy of a photon of blue light with a wavelength of 400 nm?
>
> **Solution**
>
> *Note: The appendix contains the answer.*

4.3.03 Light and Matter Interactions

Understanding the interaction between light (ie, electromagnetic radiation) and matter is critically important in scientific study. Without the reflection and collection of light, scientific measurements cannot be made. One of the many ways that light can interact with matter is through **absorption** and **emission**, which has many medical applications, such as medical imaging and radiation therapy.

Absorption occurs when a material absorbs light energy, causing the electrons in its atoms to become excited and move to higher available energy levels, as shown in Figure 4.37. (Recall from General Chemistry that electrons exist in quantized orbitals around the nucleus).

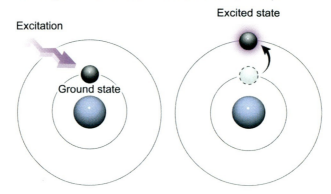

Figure 4.37 A ground state electron absorbs incoming light and is excited to higher atomic orbitals.

The amount of energy absorbed depends on the frequency of the light, which is determined by its wavelength. Different materials have different absorption spectra, which describe the range of wavelengths that they can absorb. In medical applications, absorption is used in techniques such as X-ray absorption spectroscopy, where X-rays are absorbed by materials to determine their chemical composition.

Emission occurs when a material emits light energy after being excited, either by absorbing light or through another process (Figure 4.38).

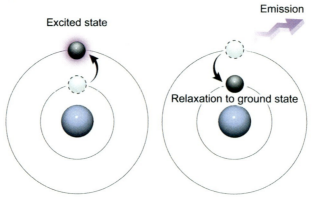

Figure 4.38 An excited electron emits a photon and falls back to a lower energy orbital.

The emitted light can have a different wavelength than the absorbed light, and the difference in energy between the two is known as the energy gap. In medical applications, emission is used in techniques such as fluorescence imaging, which uses fluorescent dyes to visualize biological tissues.

The **photoelectric effect** describes the ejection of electrons from a substance due to the **absorption** of electromagnetic radiation. Although higher-intensity (ie, higher photon count) electromagnetic radiation exhibits greater electric field oscillations, some types of electromagnetic radiation do not cause the ejection of electrons in any case, regardless of intensity. This characteristic is due to the electric potential energy between positive charges (protons) and negative charges (electrons).

For each electron, a discrete value of electric potential exists known as the **work function** W, and the magnitude of energy absorbed through electromagnetic radiation must exceed W for the ejection of an electron to occur (Figure 4.39).

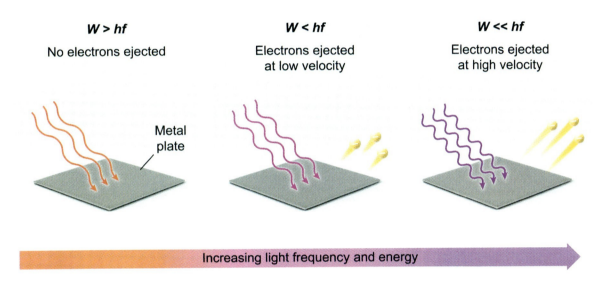

Figure 4.39 The value of the work function determines if electrons are ejected when light is incident on a metal plate.

As shown previously, the energy of electromagnetic radiation E is proportional to its frequency f via Planck's constant ($h = 6.62 \times 10^{-34}$ m²·kg/s):

$$E = hf$$

Accordingly, the ejection of an electron from a surface by incident electromagnetic radiation *depends on the frequency* of the incident electromagnetic radiation. Furthermore, the conservation of energy dictates that the energy absorbed by an electron must contribute to either the ejection of the electron (overcoming the work function) or to the kinetic energy of the electron following ejection:

$$hf = \frac{1}{2}mv^2 + W$$

Consequently, increasing the energy of the electromagnetic radiation (ie, frequency) beyond the value of the work function increases the kinetic energy of ejected electrons. However, the number of electrons ejected depends on the intensity of the electromagnetic radiation and is equivalent to the number of photons incident on the surface of the material.

> ☑ **Concept Check 4.13**
>
> Which of the following factors does *not* affect the kinetic energy of electrons emitted from a metal surface due to the photoelectric effect?
>
> A. The frequency of the incident light
> B. The intensity of the incident light
> C. The work function of the metal surface
> D. The number of electrons on the metal surface
>
> **Solution**
>
> Note: The appendix contains the answer.

4.3.04 Reflection and Refraction

Reflection is a phenomenon in which light that arrives at the interface between two media returns into the original media with a modified direction. Models of reflection assume that light comprises individual rays rather than dispersed wavefronts. This simplification allows for more accurate predictions about the pathway that light takes during reflection events (Figure 4.40).

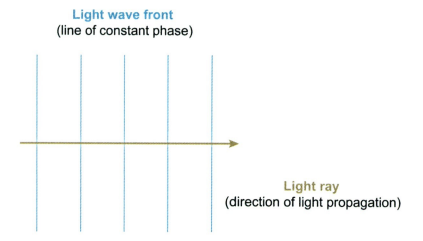

Figure 4.40 Individual rays of light traveling to the right.

Specifically, the pathway of light following a reflection event is described in terms of the angle light rays make with the **normal**, which is an imaginary line perpendicular to the interface between media. For example, the **angle of incidence** (θ_1 or θ_i) refers to the angle relative to the normal at which a ray of light approaches the interface between two media (Figure 4.41).

The principal characteristic of reflection is that the **angle of reflection** (θ_2 or θ_r) is always equal to the angle of incidence:

$$\theta_i = \theta_r$$

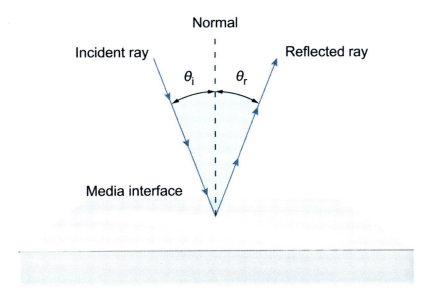

Figure 4.41 An incident ray of light is reflected when it reaches the boundary to a new medium.

When a ray of light strikes a media interface, some of the light reflects but the rest may pass through the first medium, enter the second medium, and undergo **refraction**, where its path is bent inward or outward with respect to the normal (Figure 4.42). This occurs because the speed of light changes (ie, is less than the speed of light in a vacuum) due to the new medium.

The amount of refraction is determined by each medium's refractive index n, which is the ratio of the speed of light in a vacuum c to the speed of light in the medium v:

$$n = \frac{c}{v}$$

where values of $n > 1$ represent cases in which the speed of light in that medium is less than c (ie, the speed of light in a vacuum)

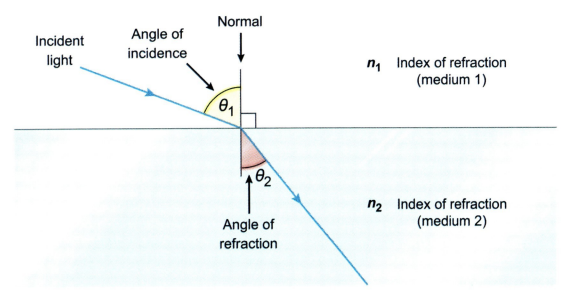

Figure 4.42 An incident light ray refracts as it enters a different medium.

Snell's law relates the index of refraction of each medium (n_1 and n_2) to the angle of incidence (θ_1) and the **angle of refraction** (θ_2), the angle at which light approaches and departs the interface between media, respectively:

$$n_1 \sin \theta_1 = n_2 \sin \theta_2$$

Rearrangement of Snell's law yields a form of the expression useful for predicting the direction of refraction when the relationship between refractive indices is known:

$$\frac{n_1}{n_2} = \frac{\sin \theta_2}{\sin \theta_1}$$

A phenomenon known as **total internal reflection** occurs when light in a medium with a high index of refraction encounters the boundary of a medium with a lower index of refraction (ie, $n_1 > n_2$) and is reflected entirely back into the first medium (Figure 4.43). This reflection occurs when the angle of incidence is greater than the **critical angle** θ_c, which is given by:

$$\theta_c = \sin^{-1}\left(\frac{n_2}{n_1}\right)$$

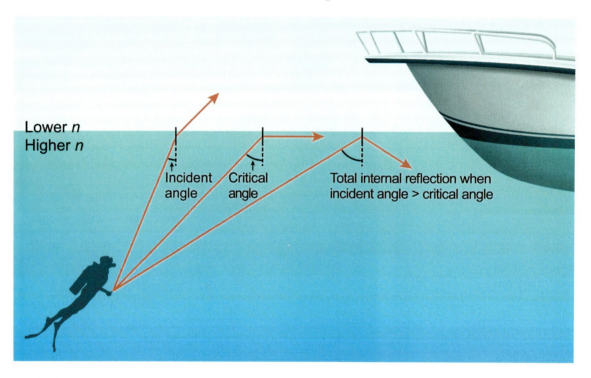

Figure 4.43 At the critical angle and greater, light rays are totally reflected back into the water and no light is refracted.

Total internal reflection is used in a variety of applications (eg, fiber optics) and allows information to be transmitted long distances without total degradation of the signal.

Another phenomenon involving reflection and refraction of light is **dispersion**, in which different colors of light are refracted by different amounts, causing them to separate (Figure 4.44).

Chapter 4: Waves, Sound, and Light

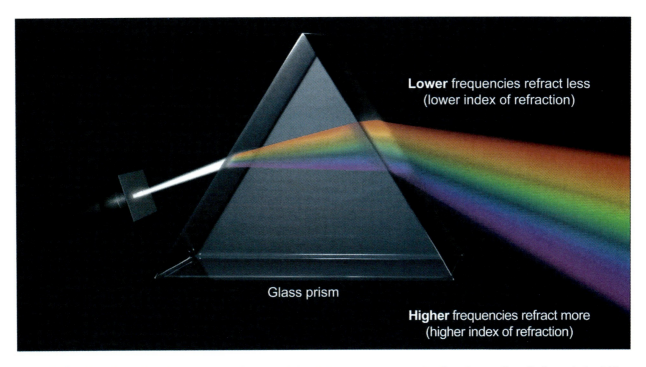

Figure 4.44 Light through a glass prism is dispersed due to the varying speeds of each wavelength (ie, color) within the glass.

Dispersion occurs because the index of refraction of a material depends on the wavelength of light, with shorter wavelengths (higher frequencies) refracted more than longer wavelengths (lower frequencies). Consequently, red light is refracted less than violet light. This effect can be seen in rainbows and in prisms, which separate white light into its component colors.

> **Concept Check 4.14**
>
> A ray of uniform light traveling in air ($n = 1.00$) strikes a flat surface of water ($n = 1.33$) with an angle of incidence of 30°. Determine the angle of refraction for the light ray.
>
> **Solution**
>
> *Note: The appendix contains the answer.*

4.3.05 Light Polarization

Polarization refers to the alignment of transverse wave oscillations in a particular orientation within the **xyz-coordinate system**. For example, a transverse wave traveling along the x-axis causes oscillations that may occur exclusively along the y- or z-axes, or at some angle between the y- and z-axes (Figure 4.45).

Polarization is unique to transverse waves because only transverse waves cause oscillations perpendicular to the direction of propagation. Consequently, electromagnetic radiation can be polarized but sound cannot.

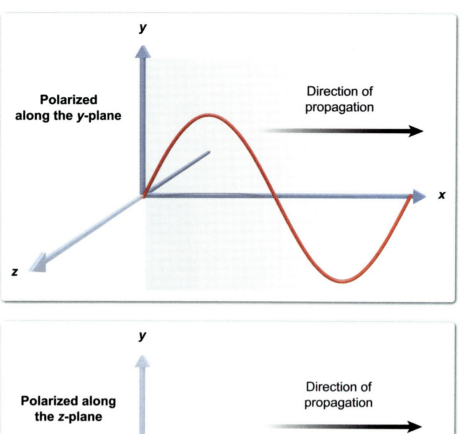

Figure 4.45 Light polarized along the *y*-plane (top) and along the *z*-plane (bottom) while traveling in the *x*-direction.

Polarization filters are optical devices that reorient the polarization of electromagnetic radiation so that radiation exiting the filter is polarized in only one orientation. For example, a linear polarization filter allows for the transmission of electromagnetic radiation oriented parallel to the **axis of polarization** but inhibits the passage of radiation oriented perpendicular to this axis (Figure 4.46).

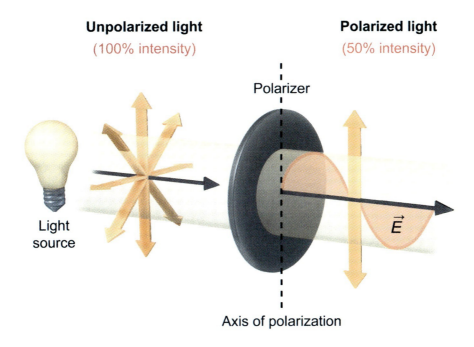

Figure 4.46 Unpolarized light can be selectively polarized by passing it through a polarizer.

Most light sources emit unpolarized light comprising waveforms oriented in all directions. Even for unpolarized light, approximately half the intensity of electromagnetic radiation is polarized along one of two perpendicular axes. Because only electromagnetic radiation with an electric field oriented parallel to the axis of polarization passes through a linear polarization filter, the total intensity of unpolarized light decreases by 50%.

The combination of two polarized waves generates a new wave with a polarization different than the two constituent waves. For example, a wave polarized along the *x*-axis totally in phase with a wave polarized along the *y*-axis produces a combined wave polarized between the *x*- and *y*-axes at a 45° angle (Figure 4.47).

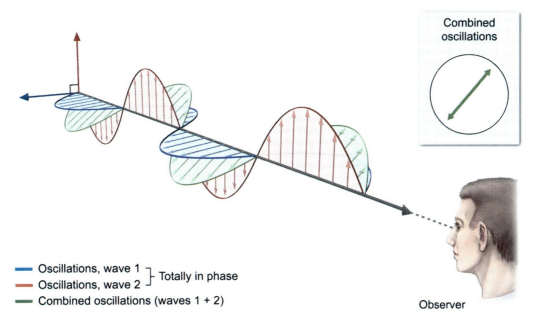

Figure 4.47 The combination of two in-phase polarized waves produces a new wave with a new polarization.

When the polarization of a combined wave **rotates** around the axis of propagation, the wave is said to be circularly polarized. **Circular polarization** occurs when two wave forms with **equal amplitude** and **perpendicular linear polarization** (ie, polarized along different axes) propagate **90° out of phase** to one another (ie, one waveform yields zero displacement when the other waveform yields a peak or a trough) (Figure 4.48).

The mismatch in phase causes each contributing wave to generate momentary oscillations that are different at every point in time, resulting in a continuous rotation in the orientation of the combined wave. If the polarization of the combined wave rotates in a clockwise direction when viewed facing the direction of propagation, the waveform is right-polarized; if the polarization rotates in a counterclockwise direction, the waveform is left-polarized.

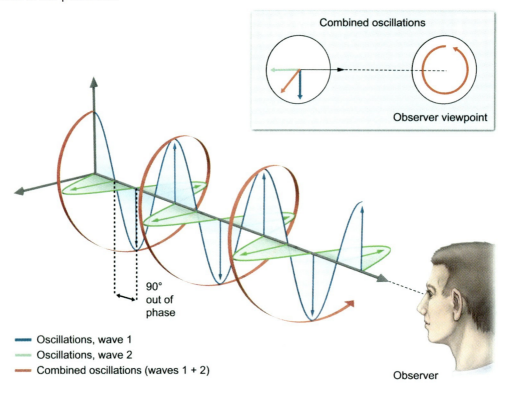

Figure 4.48 Circular polarization of light through the combination of two waveforms.

✓ Concept Check 4.15

What is unpolarized light?

A. Light waves in which the electric field vector vibrates in a single plane perpendicular to the direction of wave propagation.

B. Light waves in which the electric field vector rotates in a circle perpendicular to the direction of wave propagation.

C. Light waves in which the electric field vector vibrates in all possible directions perpendicular to the direction of wave propagation.

D. Light waves in which the electric field vector traces out an ellipse perpendicular to the direction of wave propagation.

Solution

Note: The appendix contains the answer.

4.3.06 Light Interference and Diffraction

Similar to other types of waves discussed previously, light waves experience constructive and destructive wave interference. When the peaks and troughs of two light waves overlap, they undergo **constructive interference**, forming especially bright regions (or bands). Alternatively, when the two light waves **interfere destructively**, the peaks of one wave overlap with the troughs of the other wave. Areas of destructive interference are perceived as regions (or bands) of darkness, with no light.

One such example of interference is **thin-film interference**, which is a phenomenon that occurs when light waves interact with a thin layer of transparent material, such as a soap bubble or a film of oil on water, and the reflected and transmitted light waves interfere with each other (Figure 4.49).

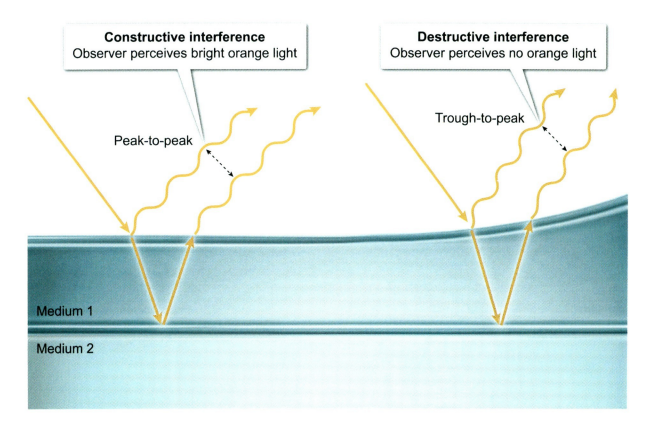

Figure 4.49 Thin-film interference of light can be constructive or destructive, depending on the film thickness or refractive index.

The resulting interference pattern depends on the thickness and refractive index of the film as well as the wavelength of the light and can be used to determine properties of the film such as its thickness or refractive index. This technique is important in medical imaging because it allows the visualization of structures within biological tissues.

Another phenomenon involving interference is **diffraction**, which is broadly defined as the bending of light around edges or objects. One such case is the classic **single-slit diffraction**, which occurs as a uniform wavefront of monochromatic light arrives at a slit (ie, confined gap) in which the **width *a* of the slit** is comparable to the **wavelength λ of the light** and results in the dispersed propagation of light away from the center of the slit.

Diffraction through a single slit produces a pattern of **dark and bright bands** on a flat background surface distant to the slit (Figure 4.50). Bands occur when the bending of light causes the distance (path length) traveled by some waves to be longer than that of others.

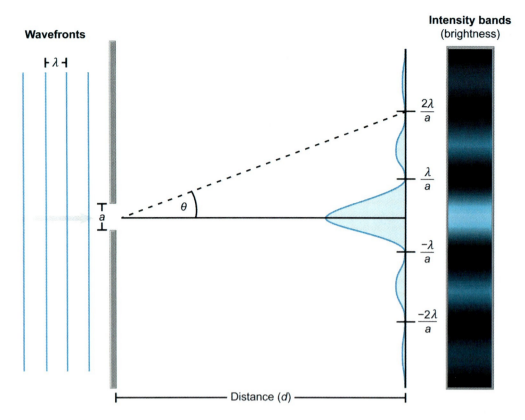

Figure 4.50 Single-slit diffraction.

When the path lengths of two light waves differ by $\lambda/2$, the waves arrive at the background **out of phase** (ie, the peak of one waveform coincides with a trough of the other). As a result, destructive interference occurs such that dark bands in the diffraction pattern are produced (Figure 4.51). Conversely, light bands result from zero or minimal destructive interference.

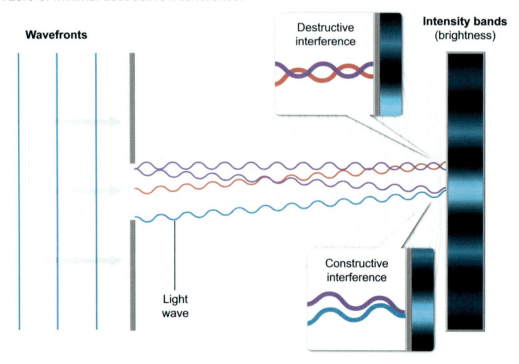

Figure 4.51 Constructive and destructive interference patterns of light from single slit diffraction.

Dark bands (minimum light intensity) due to destructive interference occur when diffracted light waves pass through the slit at angles (θ) that are a function of the slit width and an integer multiple ($m = \pm 1, \pm 2, \pm 3 \ldots$) of the wavelength:

$$\sin \theta = \frac{m\lambda}{a}$$

According to this relationship, decreased slit width or increased wavelength tends to widen the band pattern such that greater distances are present between light and dark bands. Conversely, increased slit width or decreased wavelength reduces the distance between the light and dark bands.

In a similar case, **double-slit diffraction** involves light passing through two closely spaced parallel slits, which also creates an interference pattern. Young's **double-slit experiment** shows that light behaves as both a wave and a particle, depending on how it is observed. The interference pattern produced by the double-slit experiment demonstrates the wave-like behavior of light whereas the detection of individual photons on the screen indicates the particle-like nature of light (Figure 4.52).

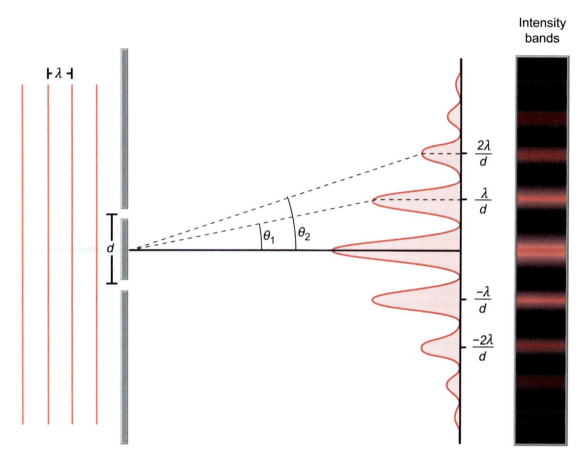

Figure 4.52 Interference pattern created by double-slit diffraction.

A **diffraction grate** is a diffractive array composed of **numerous slits** with uniform widths similar to the wavelength of incident light. Like the band patterns produced through single-slit diffraction and double-slit diffraction, dark bands that result from light passing through a diffraction grate are caused by destructive interference between light waves. Conversely, light bands are caused by constructive interference between light waves.

Compared to other diffractive phenomena, the band pattern caused by grated diffraction is characterized by relatively sharp and narrow peaks (ie, thin peaks with highly variable amplitude). This band pattern occurs because constructive and destructive interference occurs not only between waves emerging from adjacent slits (eg, double-slit diffraction) but also between waves emerging from slits separated from each other by other slits (Figure 4.53).

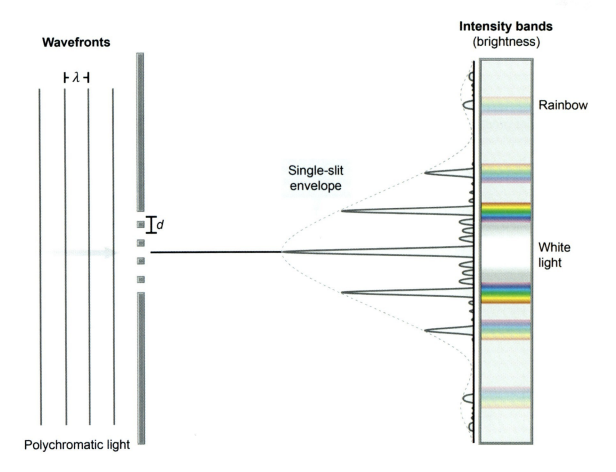

Figure 4.53 The grated diffraction of light yields patterns similar to that of single-slit and double-slit diffraction.

Light bands (regions of maximum light intensity) from the diffraction grating occur when diffracted light waves pass through the slits at angles (θ) that are a function of the distance separating each slit (d) and an integer multiple ($m = \pm1, \pm2, \pm3...$) of the wavelength:

$$\sin \theta = \frac{m\lambda}{d}$$

The sharp, narrow peaks of the band pattern produced by a diffraction grating are useful for analyzing the properties of the light that enters the slit. For example, long-wavelength light diffracts more than short-wavelength light. Consequently, transmitting **polychromatic light** (eg, white light) composed of light waves with different wavelengths through a diffraction grate produces a series of multicolored arrays (eg, rainbows) that enables an analysis of wavefront composition.

Not all forms of light undergo diffraction when incident upon one or more slits. For example, **X-rays** are a form of electromagnetic radiation that cannot undergo classic slit diffraction because the wavelengths of X-rays range from 1×10^{-11} m to 10×10^{-11} m, which are exceeded by the widths of industrially fabricated slits.

However, the wavelength of some X-rays is comparable to the typical distance between atoms within the molecular structure of most materials. Consequently, exposing a sample of a **purified and crystallized material** to X-ray radiation can produce a diffraction pattern unique to that particular substance.

For example, when two X-ray waveforms approach and depart a crystalline lattice at identical incident angles, the waveforms can constructively interfere such that a bright spot is observed on a piece of film distant to the material (Figure 4.54).

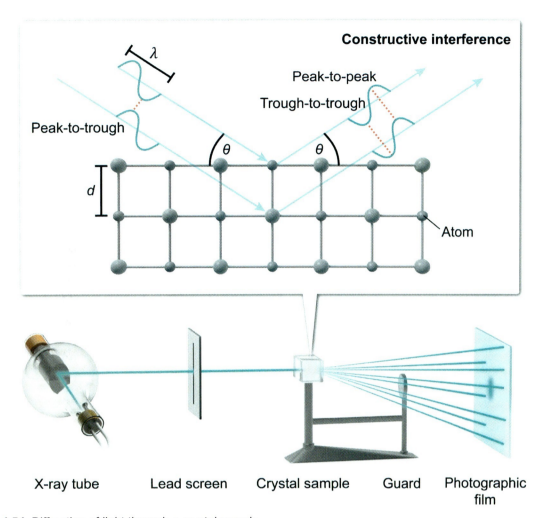

Figure 4.54 Diffraction of light through a crystal sample.

The occurrence of constructive interference is a function of the incident angle (θ), the distance between atoms (d), and an integer multiple ($m = \pm 1, \pm 2, \pm 3 \ldots$) of the X-ray wavelength (λ):

$$2d \cdot \sin \theta = m\lambda$$

Because the crystalline lattice of many materials exceeds the complexity of simple crystals, the pattern of bright spots produced on the photographic film is indicative of a molecule's three-dimensional structure and its **packing orientation** within the crystal lattice.

✓ Concept Check 4.16

A light ray passes through a single slit of width a, and a diffraction pattern is produced on a screen a distance d behind the open slit. If the slit is replaced by another slit of twice the width, how is the diffraction pattern affected?

Solution

Note: The appendix contains the answer.

4.3.07 Doppler Effect with Light

When relative motion exists between a wave source and an observer, the observed wave's frequency and wavelength differ from those of the original. This effect is known as the Doppler effect (see Concept 4.2.04) and exists for light waves as well as sound waves.

If a source of emitted light moves toward an observer, the wavelength appears shorter than the original wavelength (known as blueshift). If the source moves away from the observer, the wavelength appears longer (known as redshift).

For example, consider an astronomer observing a nearby galaxy. If the galaxy is moving toward Earth, it is viewed as a color that is bluer than its true color. If the galaxy is moving away from Earth, it appears as a redder color (Figure 4.55).

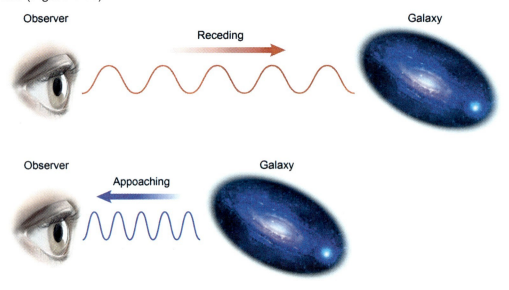

Figure 4.55 Doppler effect for astronomical objects.

In medicine, the Doppler effect has applications such as laser vibrometers, which can be used to measure respiratory rate by reflecting lasers off a patient's chest.

 Concept Check 4.17

Police officers often employ "speed guns" that use Doppler radar to perform speed measurements. A narrow beam is sent out and bounces off a target before returning to the gun. If the target is moving, the received frequency is higher or lower than the original emitted frequency, depending on whether the object is moving toward or away from the gun.

The speed of the target can be determined from the measured difference in the emitted and received frequency Δf, the emitted frequency f, and the speed of light in a vacuum c:

$$v = \frac{\Delta f}{f} \cdot \frac{c}{2}$$

If the speed gun emits a radio signal at a frequency of 10 GHz toward a car moving toward the gun at 30 m/s, what is the frequency of the reflected signal detected by the gun?

Solution

Note: The appendix contains the answer.

Lesson 4.4
Optical Instruments

Introduction

This lesson describes how the principles of geometric optics can be applied to determine the nature of images created by mirrors and thin lenses. Images of objects are created by the reflection of light rays off the surface of mirrors. When the surface of the mirror is spherical, reflected light rays either converge at a focal point or diverge away from the focal point, depending on the curvature of the mirror. This behavior leads to the formation of both real and virtual images depending on the type of mirror and location of the object.

Real and virtual images are also created by the refraction of light rays by thin lenses. The location and size of these images is determined by the thin lens equation. Furthermore, systems with multiple lenses are often used to construct optical instruments such as microscopes and telescopes. The lesson concludes with a discussion of optics applied to the human eye, particularly the correction of nearsightedness and farsightedness.

4.4.01 Reflection from Mirrors

In everyday life, the most familiar image created by the reflection of light is a **virtual image** formed by a **plane mirror**. Light shining on an object reflects off the mirror according to the law of reflection discussed in Concept 4.3.04 and enters the eye of an observer, as shown in Figure 4.58. Note that the observer and the object are on the same (reflective) side of the mirror, but the brain perceives the source of the reflected light as being behind the mirror. Hence, the object appears to be located behind the mirror.

This type of image is virtual because the actual light forming the image does not originate behind the mirror. Virtual images are always **upright** (ie, the image does not appear to be upside down with respect to the object). For a plane mirror, the image is the same size as the object. Furthermore, the distance to the image behind the mirror i is always equal to the object's distance from the mirror o:

$$i = o$$

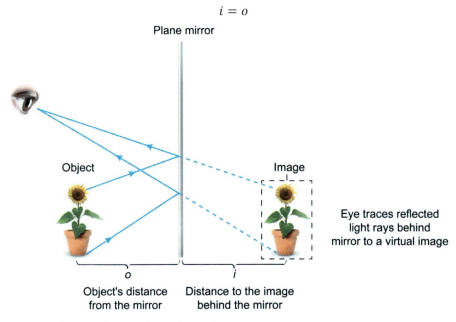

Figure 4.58 Formation of a virtual image by a plane mirror.

Mirrors with curved surfaces are also important in many applications. The most common type of curved mirror is formed by a portion of a spherical surface. The principal axis of a spherical mirror is the line that passes through its center and is perpendicular to its surface. When the radius of the sphere is large (so that the curvature of the mirror is small), these **spherical mirrors** have the property that all incoming parallel light rays are reflected to a single **focal point** located on the principal axis (Figure 4.59).

Spherical mirrors are classified as either concave or convex, depending on how the surface is curved relative to the observer. The edges of **concave (or converging) mirrors** are bent toward the observer and cause parallel light rays to converge at a focal point located in front of the mirror. In contrast, the edges of **convex (or diverging) mirrors** are curved away from the observer. With convex mirrors, parallel light rays spread out when reflected away from the mirror, but they can be traced back to a focal point, which is located behind the mirror.

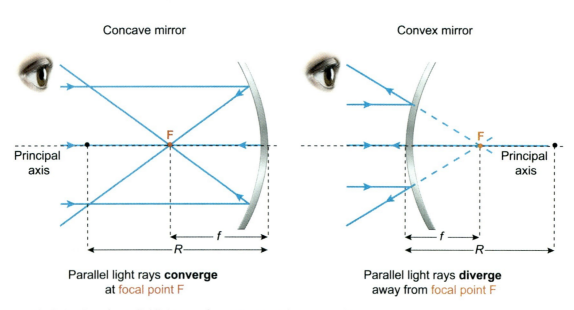

Figure 4.59 Behavior of parallel light rays for concave and convex mirrors.

The distance of the focal point from the mirror (known as the **focal length** f) is half the **radius of curvature** R of the spherical mirror surface:

$$f = \frac{R}{2}$$

For a concave mirror, f and R are positive by convention whereas f and R are negative for convex mirrors.

Concave mirrors create either virtual or **real images**, depending on the location of the object relative to the focal point. A real image is formed at the location where light rays actually converge; hence, a real image can be displayed on a screen at the location where it is formed. Unlike virtual images, **real images are inverted** (ie, the image appears upside down relative to the object). A real image is always located on the observer's (or object's) side of the mirror.

The location and size of the image (and hence the image type) can be determined by **ray tracing**. In this procedure, the object sits on the mirror's principal axis and two light rays are drawn toward the mirror starting from the top of the object.

- **Parallel ray**: The first ray travels from the tip of the object (ie, the height) parallel to the principal axis and reflects through the focal point.
- **Focal ray**: The second ray travels from the tip of the object and through the focal point. It is always reflected parallel to the principal axis.

If the reflected rays converge at a point, then the top of a real image is located at the site of the convergence. The distance of the convergence point from the principal axis is the **height of the image**. For example, when the object is *outside the focal point* of a concave mirror, a real inverted image is formed and the image is smaller than the object (Figure 4.60).

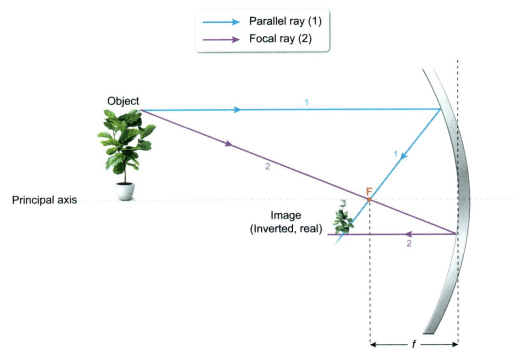

Figure 4.60 A real image created by a concave mirror. Ray 1 is the parallel ray, and ray 2 is the focal ray.

When the object is *inside the focal point*, the reflected light rays do not converge in front of the mirror but instead diverge (Figure 4.61). However, when the reflected rays are traced back behind the mirror, they do converge at a single point and produce a virtual image that is larger than the object.

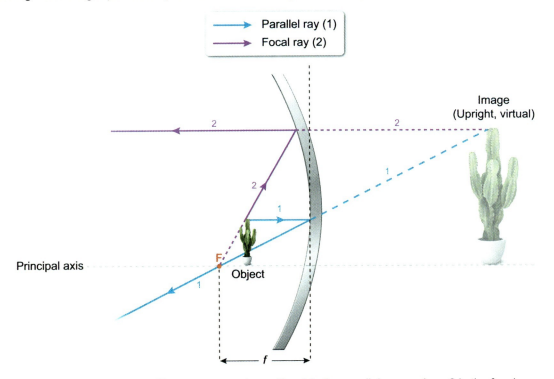

Figure 4.61 A virtual image created by a concave mirror. Ray 1 is the parallel ray, and ray 2 is the focal ray.

Although the type of image formed by a concave mirror depends on the location of the object, *convex mirrors always create virtual images* that appear behind the mirror. An example of image formation for a convex mirror is shown in Figure 4.62.

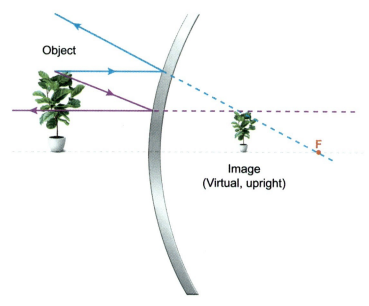

Figure 4.62 Example of a virtual image created by a convex mirror. Ray 1 is the parallel ray, and ray 2 is the focal ray.

> ☑ **Concept Check 4.18**
>
> A person in a department store looks into a spherical mirror near the ceiling and sees a smaller image of themself. What is the type of mirror that this person sees?
>
> **Solution**
>
> *Note: The appendix contains the answer.*

> ## ✓ Concept Check 4.19
>
> An object (represented schematically by the arrow) is placed in front of a convex mirror, as shown in the figure. Perform ray tracing to determine the location and size of the image produced by the mirror.
>
>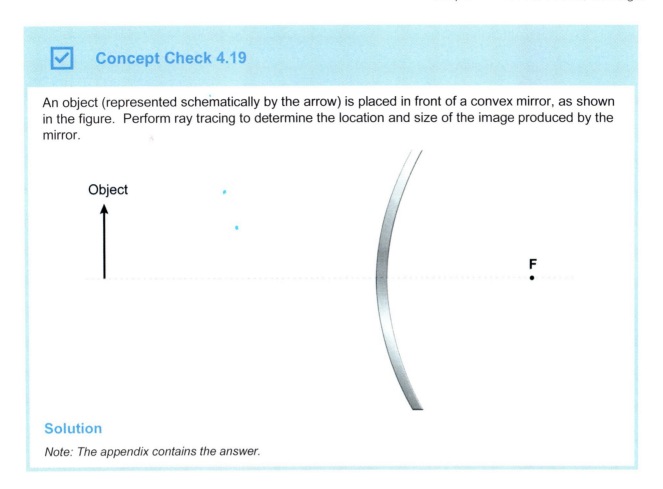
>
> ### Solution
> Note: The appendix contains the answer.

4.4.02 Refraction from Thin Lenses

Lenses are a key component of a wide variety of optical instruments, such as cameras, eyeglasses, microscopes, and telescopes. Lenses form virtual and real images of objects using the **law of refraction** discussed in Concept 4.3.04. When light rays travel from one medium into a different medium, they are bent toward or away from the normal direction according to Snell's law, which depends on the values of the indices of refraction n_1 and n_2 of both media.

A thin lens is formed by a piece of glass or plastic material with both sides gently curved into either a concave or convex shape like a spherical mirror. A **convex (or converging) lens** is shaped so that the center of the lens is thicker than the edges. As a result, parallel rays of light passing through a convex lens are refracted such that they bend toward each other, converging at a single focal point, as shown in Figure 4.63.

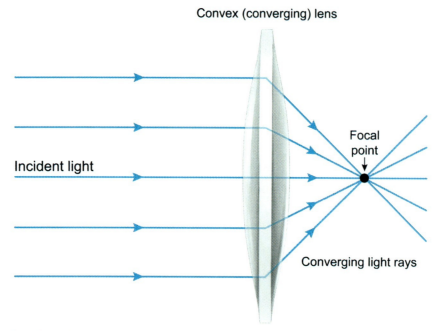

Figure 4.63 Behavior of parallel light rays passing through a convex (converging) lens.

Conversely, a **concave (or diverging) lens** is thinner at the center than at the edges, causing parallel rays of light passing through the lens to be bent away from each other (see Figure 4.64). Just as for diverging spherical mirrors, the refracted light rays diverging from a concave lens can be traced back to a focal point in front of the lens.

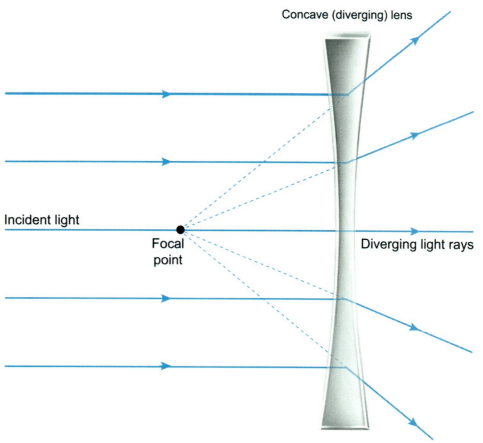

Figure 4.64 Behavior of parallel light rays entering a concave (diverging) lens.

Note that lenses have **two focal points** (one on each side equal distances from the lens) because they are formed by two curved surfaces. Light rays converge (or appear to converge) at one of the focal points, depending on the type of lens and which surface the incident light strikes.

Lenses form virtual and real images closely analogous to the spherical mirrors described in Concept 4.4.01 however, with some important differences. Because refracted light passes through the lens (instead of being reflected, as with a mirror), the observer and the object must be on *different sides* of the lens.

Mirrors and lenses of the same shape have different effects on light rays. A concave mirror causes rays to converge, but a concave lens causes rays to diverge. Similarly, a convex mirror causes light to diverge, but a convex lens causes light to converge. For a summary of these properties, see Figure 4.65.

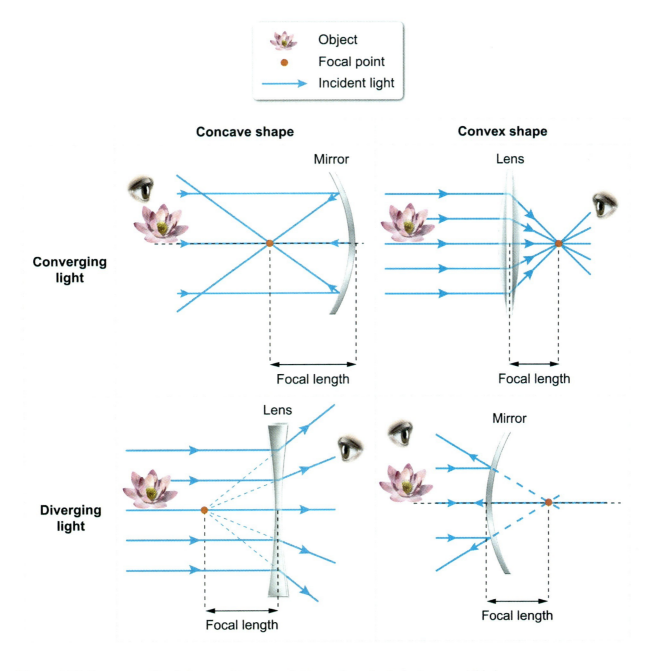

Figure 4.65 Summary of the behavior of incoming light rays for spherical mirrors and thin lenses.

The location of an image produced by a thin lens can be determined by ray tracing, just as for spherical mirrors.

- **Parallel ray**: The first ray travels from the tip of the object (ie, the height) parallel to the principal axis of the lens. For a converging lens, the ray is refracted through the lens' focal point located on the far side of the lens. For a diverging lens, the ray is refracted as though it originates at the focal point located on the near side of the lens.
- **Central ray**: The second ray travels from the tip of the object through the center of the lens. For both types of thin lenses, the ray is not refracted and continues in a straight line.

If the refracted rays intersect at a point opposite the observer's side of the lens, the image is real. **Real images** created by lenses are always **inverted**, just like real images created by spherical mirrors. An example of a real image created by converging lens is shown in Figure 4.66.

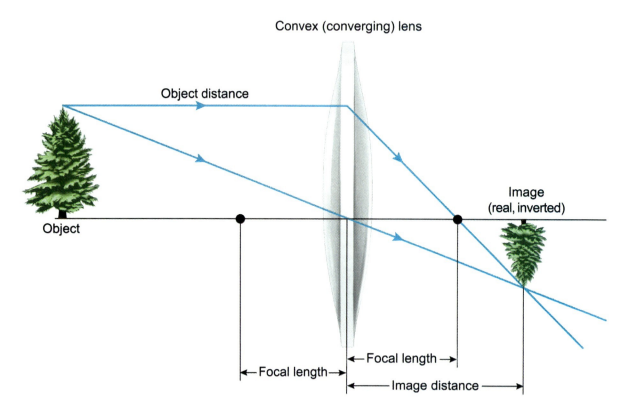

Figure 4.66 Formation of a real image by a converging lens. The object is located outside the focal length of the lens.

However, if the refracted rays do not intersect, they can be traced to a convergence point on the observer's side of the lens, and the observer sees a **virtual image**. Virtual images created by lenses are always **upright**, just like virtual images created by spherical mirrors. An example of a virtual image created by a diverging lens is shown in Figure 4.67.

Chapter 4: Waves, Sound, and Light

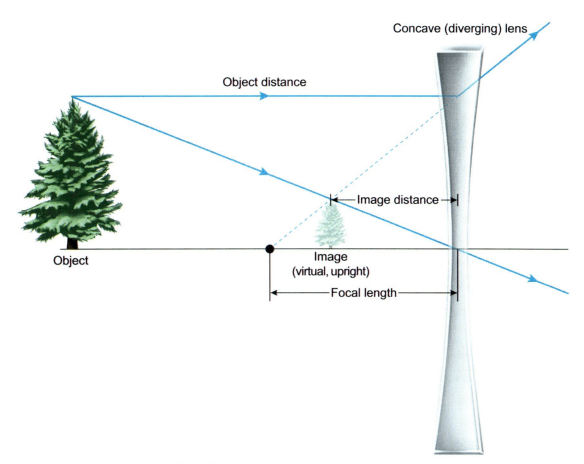

Figure 4.67 A virtual image created by a diverging lens.

✅ Concept Check 4.20

An object is placed inside the focal length of a converging lens, as shown in the figure. Use ray tracing to determine the location and size of the image created by the lens.

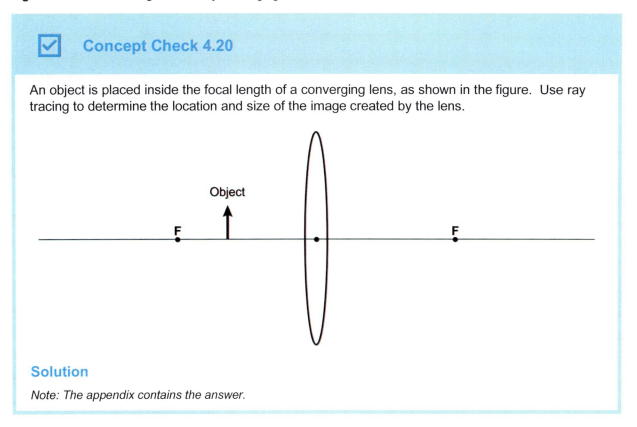

Solution

Note: The appendix contains the answer.

Although ray tracing is useful for qualitative analysis of lenses, the trigonometry associated with ray tracing can be used to derive the **thin lens equation**:

$$\frac{1}{o} + \frac{1}{i} = \frac{1}{f}$$

where o is the object's distance from the lens, i is the image distance from the lens, and f is the lens' focal length (see Figure 4.68 for an example).

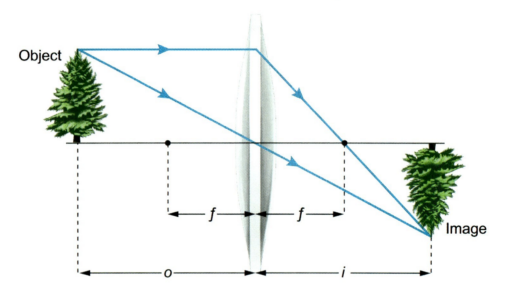

Figure 4.68 The thin lens equation.

The thin lens equation is also sometimes expressed in terms of the **lens strength (or power)** S, which is the inverse of the focal length:

$$S = \frac{1}{f} \quad \Rightarrow \quad S = \frac{1}{o} + \frac{1}{i}$$

The unit of lens strength is the **diopter** (D), which in SI units is inverse meters (m^{-1}).

The thin lens equation is an important equation to remember for the exam. Note that the thin lens equation can also be applied to the images created by spherical mirrors, but on the exam this equation is typically applied only to questions involving lenses.

Proper use of the thin lens equation requires knowledge of the **conventions for the signs** of object distance, image distance, and focal length in the equation. Object distance is positive when the object is on the opposite side of the lens relative to the observer. For the single lens systems included on the exam, the *object distance is always positive* (ie, $o > 0$).

Image distance is positive if the image forms on the same side of the lens as the observer; hence, real images have a positive image distance. Virtual images, which form on the opposite side of the lens from the observer, have a negative image distance (Table 4.3). A converging lens has a positive focal length $f > 0$, but a diverging lens has a negative focal length $f < 0$.

Chapter 4: Waves, Sound, and Light

Table 4.3 Conventions for images created by thin lenses.

Real image	Virtual image
Same side of lens as observer	Opposite side of lens as observer
Image distance $i > 0$	Image distance $i < 0$

For example, suppose an object is placed 10 mm from a converging lens with a focal length of 15 mm. Rearranging the thin lens equation yields:

$$\frac{1}{i} = \frac{1}{f} - \frac{1}{o}$$

Inserting $f = +10$ mm and $o = +15$ mm gives:

$$\frac{1}{i} = \frac{1}{15 \text{ mm}} - \frac{1}{10 \text{ mm}}$$

Finding a common denominator and solving for i yields:

$$\frac{1}{i} = \frac{2}{30 \text{ mm}} - \frac{3}{30 \text{ mm}} = \frac{-1}{30 \text{ mm}}$$

$$i = -30 \text{ mm}$$

Consequently, the image is virtual and located 30 mm from the lens on the side opposite the observer.

The size of the images created by a lens can be determined from their **magnification**. The magnification of the image M is equal to the ratio of the image height h_i to the object height h_o:

$$M = \frac{h_i}{h_o}$$

Note that $M < 0$ when the image is inverted, so real images always have a negative M value and virtual, upright images always have a positive M value ($M > 0$). Furthermore, when the absolute value of M is less than 1 ($|M| < 1$), the image is smaller than the object. For an absolute value of M greater than 1 ($|M| > 1$), the image is larger than the object.

Using simple trigonometry, the magnification can also be expressed in terms of the ratio of image distance to object distance:

$$M = \frac{h_i}{h_o} = -\frac{i}{o}$$

The negative sign appears in the equation to account for the conventions of the sign for the image distance, as described in Table 4.3.

Consider the example of the converging lens detailed previously, with object distance $o = 10o = 10$ mm and image distance $i = -30$ mm. Substituting these values into the magnification equation yields a magnification factor of $M = 3$, implying the image formed is three times as large as the object (Figure 4.69).

$$M = -\left(\frac{-30 \text{ mm}}{10 \text{ mm}}\right) = 3$$

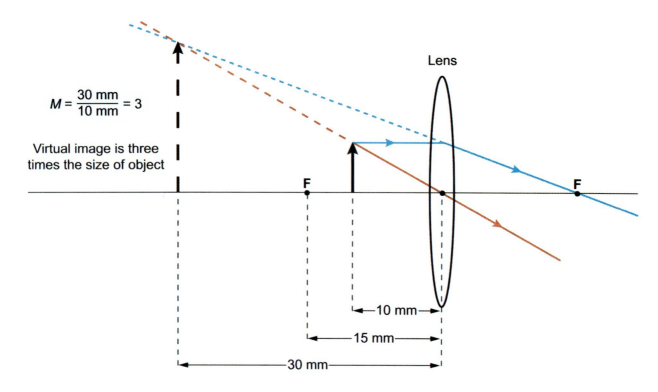

Figure 4.69 Magnification of an object placed inside the focal length of a converging lens.

> **Concept Check 4.21**
>
> An object is held 5 cm in front of a lens and creates an image at a distance of −3 cm. What is the type of lens and the lens' strength in diopters? What is the magnification of the image?
>
> **Solution**
>
> *Note: The appendix contains the answer.*

4.4.03 Combinations of Lenses and Lens Aberration

Many optical instruments, such as microscopes and telescopes, consist of systems of **multiple lenses** placed in sequence. Just as for a single lens, the images created by these combinations (ie, not in direct contact) of lenses can be determined from ray tracing and the thin lens equation. For example, the image created by the first lens in a sequence becomes the object for the second lens (Figure 4.70).

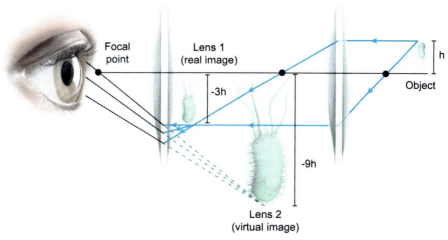

Figure 4.70 Image formed by two converging lenses.

The details of image formation through multiple lenses are beyond the scope of the exam, yet questions involving the **total magnification** and the **total strength (or power)** of a system of lenses may appear. For multiple lenses placed in sequence, the total magnification of the lens series is a *product* of each lens' individual magnification:

$$M_{total} = M_1 \cdot M_2 \cdot M_3 \cdots M_n$$

Furthermore, total lens strength is equal to the *sum* of the individual lens strengths:

$$S_{total} = S_1 + S_2 + \cdots$$

Equivalently, in terms of focal lengths f:

$$\frac{1}{f_{total}} = \frac{1}{f_1} + \frac{1}{f_2} + \cdots$$

Recall that both magnification and strength can be either positive or negative. For example, converging (convex) lenses have positive strengths S, and therefore a combination of converging lenses has a greater strength, which is equivalent to a shorter focal length (Figure 4.71).

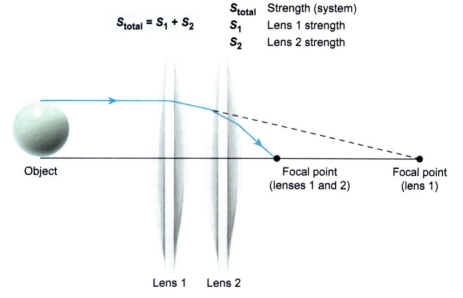

Figure 4.71 Total strength of two convex lenses in sequence.

Conversely, placing a diverging lens in series with a converging lens can decrease the total strength and produce an overall greater focal distance.

> ### ☑ Concept Check 4.22
>
> (a) A converging lens of focal length 5 cm is placed in sequence with a diverging lens of −10 cm. What is the total strength of this system?
>
> (b) A real image three times the size of the object is created by the converging lens. The virtual image created by the diverging lens is $\frac{3}{5}$ the size of the image created by the first converging lens. What is the total magnification?
>
> #### Solution
>
> *Note: The appendix contains the answer.*

In real-world applications, optical instruments are not constructed with perfectly ideal lenses or ideal spherical mirrors. The ideal lenses and mirrors discussed in this lesson always focus parallel light rays at a single point. However, lenses and mirrors are subject to **aberrations** and may not concentrate all rays at a single location. As a result, images that form from flawed lenses are not focused and appear blurry. The two most common types of aberration effects are chromatic aberration and spherical aberration.

Chromatic aberration refers to the formation of blurry images because of dispersion through a lens. As discussed in Concept 4.3.04, **dispersion** is the phenomenon of light separating into the spectrum of colors because different frequencies of light have slightly different refractive indices in the same medium. As white light consisting of all the different colors passes through a lens, violet (higher frequency) wavelengths refract more than red (lower frequency) wavelengths. Consequently, each color of light is focused at a slightly different location (Figure 4.72).

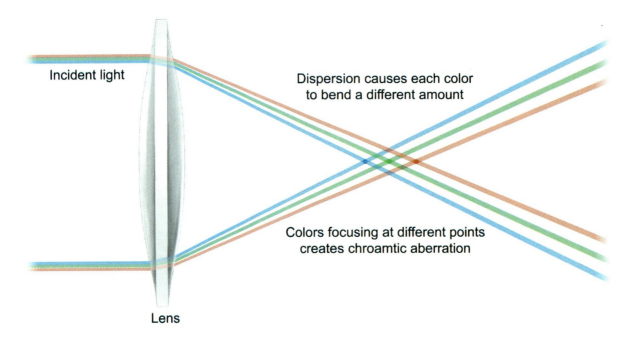

Figure 4.72 Chromatic aberration due to dispersion.

Chromatic aberration can be corrected by placing a diverging lens with a different index of refraction directly in contact with the converging lens, creating a lens couplet or doublet (see Figure 4.73).

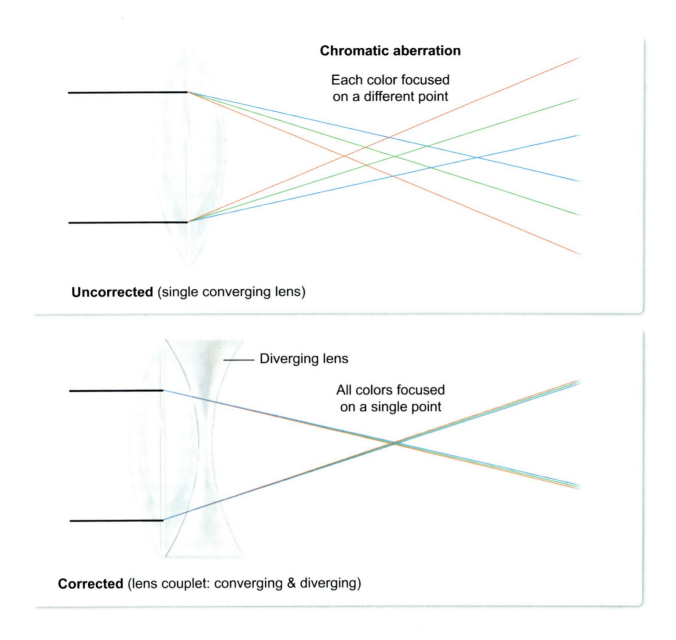

Figure 4.73 Correction of chromatic aberration by adding a diverging lens.

In contrast, **spherical aberration** occurs even when light of a single color/frequency passes through a lens. In this case, light rays passing through the lens near its edges (ie, at greater distance from the principal axis) do not converge at a single focal point. The spherical aberration in a converging lens can be corrected by using an aspherical lens in which the thickness of the lens periphery is *reduced* relative to a perfectly rounded lens (Figure 4.74).

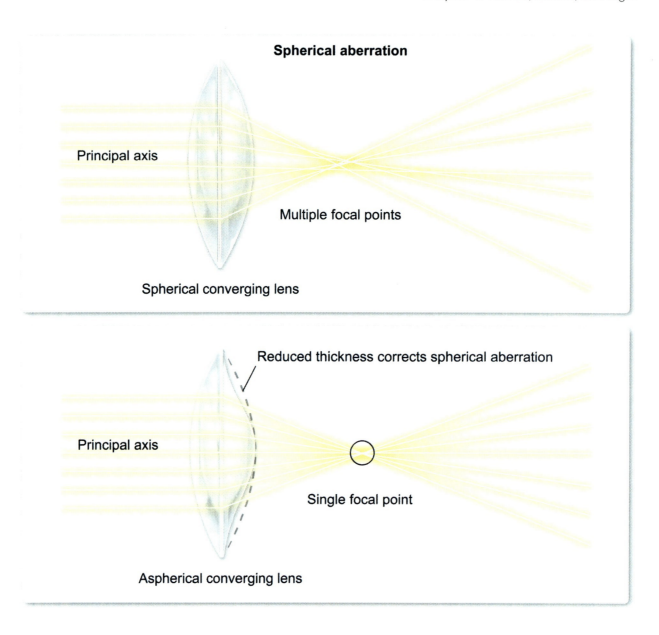

Figure 4.74 Spherical aberration for a converging lens.

4.4.04 The Human Eye

The **eye** is a biological optical organ responsible for vision. The primary components of the eye are the **cornea** and **lens**, both of which form a converging lens that focuses incoming light rays onto the **retina**, a structure analogous to the film or sensor within a camera (see Figure 4.75).

Chapter 4: Waves, Sound, and Light

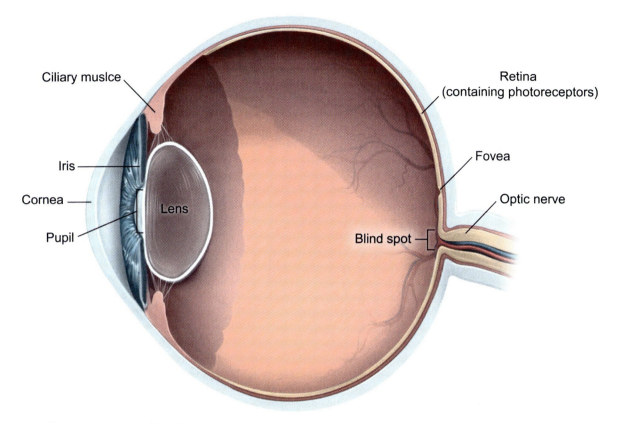

Figure 4.75 The structure of the human eye.

The images created on the retina are real and therefore inverted, but the brain inverts the images again so that objects appear upright.

For objects to appear clear and focused, images must form exactly on the retina. Consequently, the image distance *i* should always equal the lens-to-retina distance, which is about 2.0 cm (see Figure 4.76).

$$S = \frac{1}{o} + \frac{1}{i}$$

- **S**   Lens strength
- **o**   Object distance
- **i**   Image distance

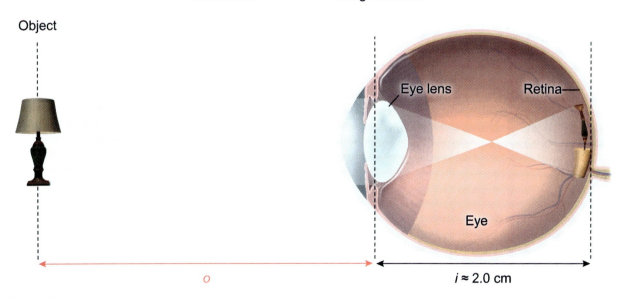

Figure 4.76 Image formation in the eye.

233

The thin lens equation implies that as the object distance o changes, the strength S (or equivalently, the focal length) of the lens also changes. When viewing distant objects (ie, where $o \approx \infty$) the ciliary muscles relax, which keeps the focal length of the lens about 2.0 cm, or equivalently in terms of strength:

$$S \approx \frac{1}{2.0 \text{ cm}}\left(\frac{100 \text{ cm}}{1 \text{ m}}\right) \approx 50 \text{ D}$$

To view closer objects, the ciliary muscles act to make the lens thicker, *increasing S* and decreasing the focal length so that objects remain properly focused on the retina.

> ☑ **Concept Check 4.23**
>
> A person can increase the strength of the lens in their eye to a maximum of 54 D to clearly view a nearby object. What is the minimum distance between the object and the eye?
>
> **Solution**
>
> *Note: The appendix contains the answer.*

Two types of vision problems are common, in which objects appear blurry and unfocused at certain distances. **Nearsightedness (or myopia)** refers to vision where nearby objects are clear but objects far away are blurry. Nearsightedness arises because the lens is unable to focus distant objects on the retina, and instead the image is focused in front of the retina (see Figure 4.77).

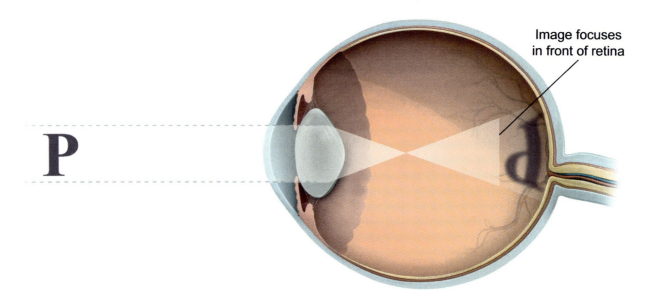

Figure 4.77 Nearsightedness (myopia).

Nearsightedness is corrected using eyeglasses with a diverging lens, which creates a lens system with a *reduced strength* and allows the image to properly appear on the retina (see Figure 4.78).

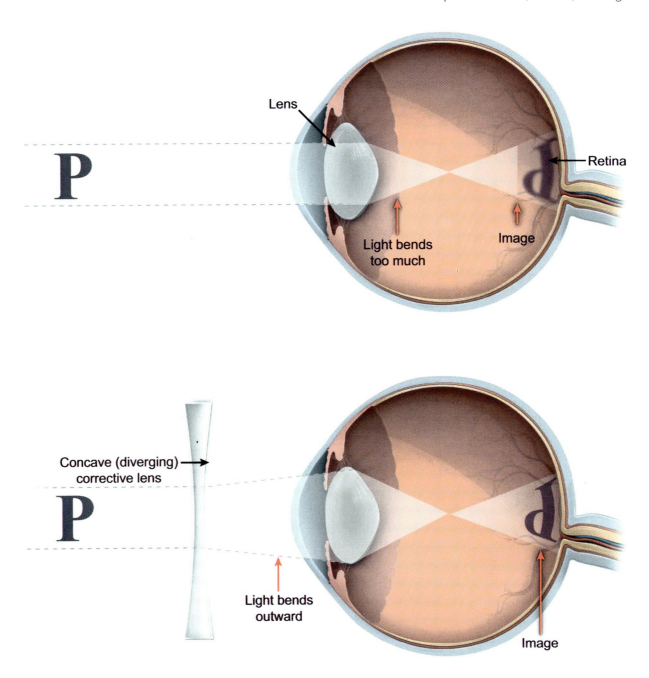

Figure 4.78 Correction of nearsightedness (myopia) with a diverging lens.

In contrast, **farsightedness (hyperopia)** is vision where distant objects are clear but nearby objects are blurry. Farsightedness arises because the lens creates an image located behind the retina (see Figure 4.79).

Chapter 4: Waves, Sound, and Light

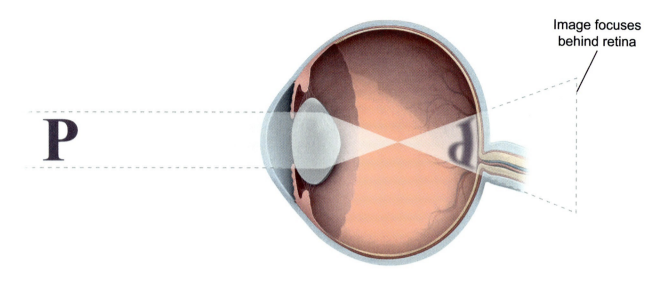

Figure 4.79 Farsightedness (hyperopia).

Farsightedness can be corrected using eyeglasses with a converging lens, which creates a lens system with a greater strength and allows the image to properly appear on the retina (Figure 4.80).

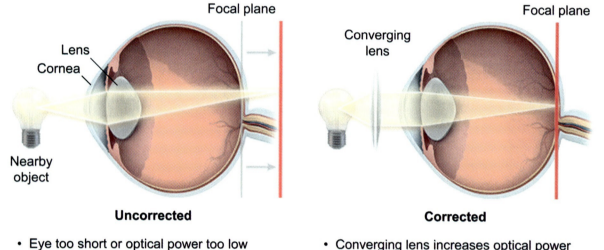

Figure 4.80 Correction of farsightedness (hyperopia) with a converging lens.

 Concept Check 4.24

A person is prescribed eyeglasses with a lens strength of −1.5 D. What vision problem does this person likely have?

Solution

Note: The appendix contains the answer.

END-OF-UNIT MCAT PRACTICE

Congratulations on completing **Unit 4: Light and Sound**.

Now you are ready to dive into MCAT-level practice tests. At UWorld, we believe students will be fully prepared to ace the MCAT when they practice with high-quality questions in a realistic testing environment.

The UWorld Qbank will test you on questions that are fully representative of the AAMC MCAT syllabus. In addition, our MCAT-like questions are accompanied by in-depth explanations with exceptional visual aids that will help you better retain difficult MCAT concepts.

TO START YOUR MCAT PRACTICE, PROCEED AS FOLLOWS:

1) Sign up to purchase the UWorld MCAT Qbank
   IMPORTANT: You already have access if you purchased a bundled subscription.
2) Log in to your UWorld MCAT account
3) Access the MCAT Qbank section
4) Select this unit in the Qbank
5) Create a custom practice test

Unit 5 Thermodynamics

Chapter 5 Thermodynamics and Gases

5.1 Thermodynamic Systems

    5.1.01   Open and Closed Systems
    5.1.02   State Function Definition

5.2 Thermodynamic Laws

    5.2.01   Zeroth Law and Temperature
    5.2.02   First Law of Thermodynamics
    5.2.03   Second Law and Entropy

5.3 Heat

    5.3.01   Heat Transfer Mechanisms
    5.3.02   Coefficient of Expansion
    5.3.03   Heat Capacity and Calorimetry

5.4 Phases of Matter

    5.4.01   Phase Diagrams
    5.4.02   Heat of Fusion and Heat of Vaporization

5.5 Kinetic Theory of Gases

    5.5.01   Ideal and Nonideal Gases
    5.5.02   Kinetic Theory
    5.5.03   Heat Capacity at Constant Volume and Pressure

Lesson 5.1

Thermodynamic Systems

Introduction

When two systems are at different temperatures, a transfer of thermal energy occurs between the two systems, which is known as heat. Thermodynamics is the study of this energy transfer and its relationship to the work that this heat can perform on objects. This lesson describes the basic definitions of thermodynamic systems, state functions, and path functions that characterize thermodynamic processes.

5.1.01 Open and Closed Systems

In thermodynamics, a **system** is a region of space that for investigative purposes is separated from the rest of the universe. Examples of systems include a chemical reaction in a sealed container, a room in a house, or even the entire Earth. Three types of thermodynamic systems exist—open systems, closed systems, and isolated systems—which are dependent on the type of barrier between the system and everything outside the system (see Figure 5.1).

An **open system** allows heat and matter to be exchanged with the surrounding environment. For example, a room with an open window is considered an open system because both matter (eg, air) and heat can flow in and out of the room.

A **closed system** allows heat but not matter to be exchanged with the surroundings. If the windows of a room are closed, then matter is not able to be transferred into and out of the room, but heat can still flow (via conduction, as described in Concept 5.3.01).

Finally, an **isolated system** does not allow either heat or matter to be exchanged with the surrounding universe. In this case, the room is also sealed and heavily insulated so that any transfer of heat to the surrounding environment is negligible.

Figure 5.1 Thermodynamic systems.

Chapter 5: Thermodynamics and Gases

> ☑ **Concept Check 5.1**
>
> A nuclear reaction takes place inside a chamber surrounded by water. During the reaction, scientists measure an increase in the water's temperature, but no radioactive particles are detected in the water. What type of system is the chamber?
>
> **Solution**
>
> *Note: The appendix contains the answer.*

5.1.02 State Function Definition

A system is said to be in **thermodynamic equilibrium** if no large-scale transfers of matter or energy (eg, heat) occur in the system. A state of equilibrium can be characterized by two different types of quantities: state functions and process (or path) functions.

State functions are thermodynamic variables that describe the present equilibrium state of the system without reference to details of how it arrived at that state. Hence, state functions are independent of the path taken by the system to arrive at its present state (see Figure 5.2). State functions include a system's pressure, volume, and temperature. Other state functions discussed in General Chemistry are enthalpy, Gibbs free energy, and Helmholtz free energy.

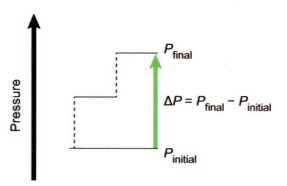

Changes in pressure are the same and do not depend on path between initial and final states

Figure 5.2 State functions such as pressure are independent of the path taken between equilibrium states.

Process functions describe the path taken by a system to transition from one equilibrium state to another. These transitions are due to a net flow of energy in the form of work or a transfer of heat. For example, the loss or gain of heat is a process function because it describes the path taken by a system from its current pressure, volume, and temperature to a different set of values (see Figure 5.3).

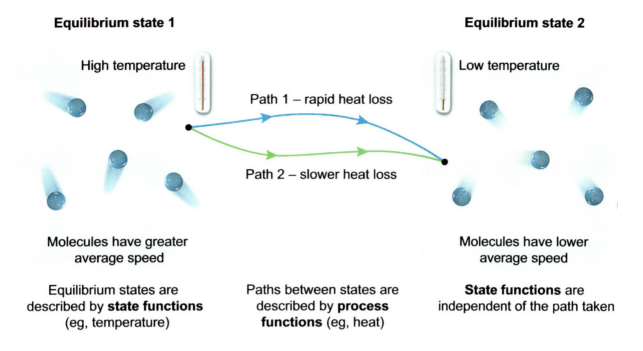

Figure 5.3 State vs. process functions.

Processes in which an important thermodynamic quantity remains constant are assigned special names (Figure 5.4). **Adiabatic processes** occur when *no heat is exchanged* between the system and the environment. In **isobaric** processes, the pressure of the system remains constant. Similarly, in **isochoric (or isovolumetric)** processes, the volume of the system does not change. Finally, **isothermal** processes occur when the temperature of the system remains constant. Lesson 5.2 introduces the concept of temperature and describes the basic thermodynamic laws.

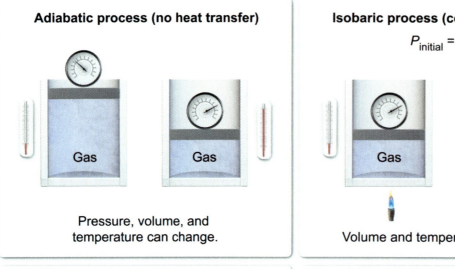
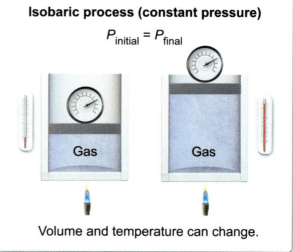
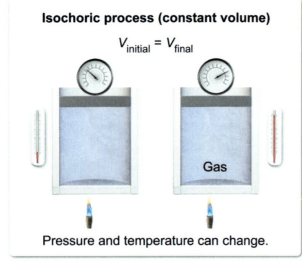
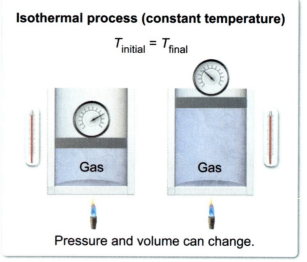

Figure 5.4 Thermodynamic processes.

> ☑ **Concept Check 5.2**
>
> The temperature and internal energy of a gas changes as a piston does 5 J of work compressing a gas to a smaller volume.
>
> (a) Which thermodynamic variable is a process function?
>
> (b) The compression of the gas by the piston could potentially be which type(s) of process?
>
> **Solution**
>
> *Note: The appendix contains the answer.*

Lesson 5.2
Thermodynamic Laws

Introduction

This lesson introduces three laws of thermodynamics, which describe all thermodynamic processes in nature. The zeroth law of thermodynamics is a fundamental law describing the notion of equilibrium and its relation to temperature. The first law of thermodynamics relates the change in a system's internal energy to heat and work. The lesson concludes by introducing the concept of entropy and the second law of thermodynamics.

5.2.01 Zeroth Law and Temperature

Temperature is a measure of how hot or cold a system is, and it is proportional to the system's internal energy. As described in Concept 5.5.02, temperature can be related to the kinetic energy of the atoms and molecules in a system. For example, in an ideal gas an increase in the temperature is associated with an increase in the average velocity of the particles in the gas.

Temperature is measured with a **thermometer**, and two common sets of temperature units appear on the exam. The **Celsius (°C)** scale is used in everyday life and in practical scientific applications. The scale is fixed by the freezing and boiling points of water under typical conditions, which are set to 0 °C and 100 °C, respectively. On the Celsius scale, normal body temperature is about 37 °C.

The **Kelvin (K)** scale is used in theoretical contexts and is fixed by **absolute zero** (0 K), which is the theoretical limit where the motion of particles in the system is a minimum. Note that the increments of the Celsius and Kelvin scales are equal. For a change in temperature ΔT:

$$\Delta T \text{ °C} = \Delta T \text{ K}$$

For example, changing the temperature of an object by 10 °C is equivalent to changing its temperature by 10 K. Hence, a temperature value in K is a linear shift of a temperature value in °C:

$$\text{K} = \text{°C} + 273.15$$

Therefore, the value of absolute zero on the Celsius scale is −273.15 °C (Figure 5.5).

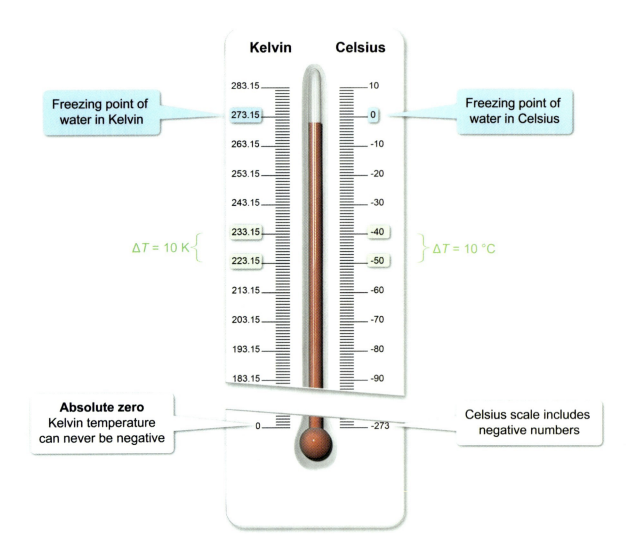

Figure 5.5 Temperature in Kelvin and Celsius.

A system in thermodynamic equilibrium has a uniform temperature value. The **zeroth law of thermodynamics** is a basic principle that relates systems in equilibrium with one another. If system A is in equilibrium with system B, and system B in equilibrium with system C, then the zeroth law implies that system A must be in equilibrium with system C (see Figure 5.6).

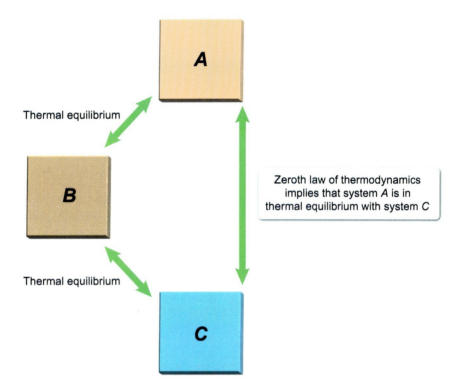

Figure 5.6 Zeroth law of thermodynamics.

In general, if two systems are at different temperatures, a flow of heat energy must exist between the two systems. Therefore, all three systems (A, B, and C) must be at the same temperature for no heat flow to exist between them (Figure 5.7).

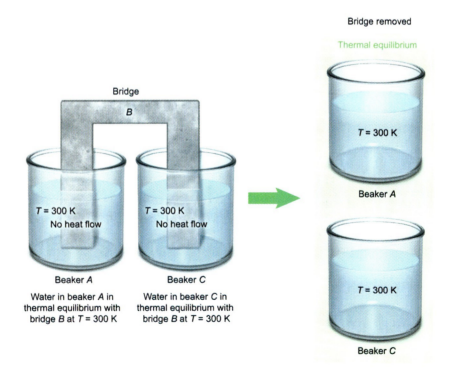

Figure 5.7 Zeroth law of thermodynamics applied to two beakers and a bridge.

> **Concept Check 5.3**
>
> An experiment cools helium to a liquid state at 3 K. What is this temperature in °C?
>
> **Solution**
>
> *Note: The appendix contains the answer.*

5.2.02 First Law of Thermodynamics

A system's internal energy U is the total energy contained in its molecules, and it is directly proportional to the system's temperature T:

$$U \propto T$$

The principle of **conservation of energy** (Concept 1.5.04) means that energy can only be transferred from one form to another; it cannot be created or destroyed. As discussed in Lesson 5.1, transfers of energy in thermodynamic systems occur via a transfer of **heat** or by **work**.

Conservation of energy can be expressed as the **first law of thermodynamics**, which states that the change in internal energy ΔU is equal to the difference between the amount of heat Q transferred to a system and the net work W done *by the system* on the surroundings:

$$\Delta U = Q - W$$

When $W > 0$, the system does net work on the surroundings and when $W < 0$, the surroundings do net work on the system. The minus sign in the first law appears because work done by the system *decreases* the system's internal energy. This convention is typically used in physics (Figure 5.8).

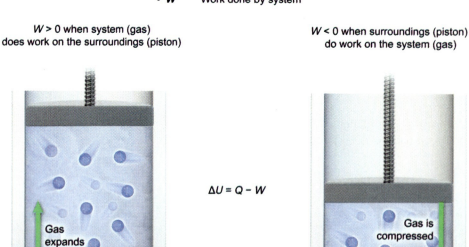

Figure 5.8 Conventions used in physics for the first law of thermodynamics.

Note that another convention exists (used in chemistry) where the first law is expressed with a plus sign:

$$\Delta U = Q + W$$

In this case, W is defined as the work done *on the system* by the surroundings. Thus, when the surroundings do net work on the system, W > 0; and when the system does net work on the surroundings, W < 0. The plus sign appears in this form of the first law because work done on the system *increases* the system's internal energy.

For heat, Q > 0 when heat flows into the system (increasing energy) and Q < 0 when heat flows out of the system (decreasing energy). For a summary of heat and work conventions see Figure 5.9.

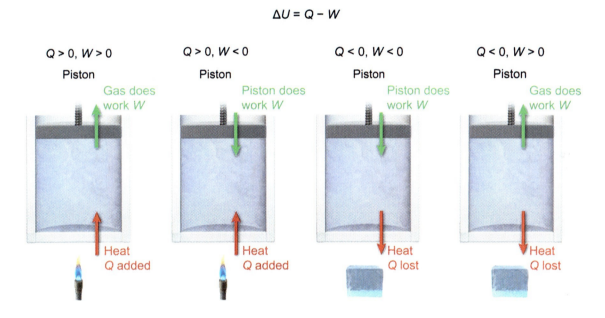

Figure 5.9 Summary of sign conventions for heat and work.

Furthermore, many applications of thermodynamics involve fluids, particularly gases. For isobaric processes (Concept 5.1.02), work can be expressed as the product of the constant pressure P and the change in volume ΔV (Concept 2.2.05):

$$W = P \cdot \Delta V$$

The overall sign of this work depends on the choice of convention discussed above. For example, suppose a piston (the surroundings) compresses a gas (the system) inside a container at a pressure of 200 Pa from an initial volume of 3 m³ to a final volume of 1 m³ (Figure 5.10).

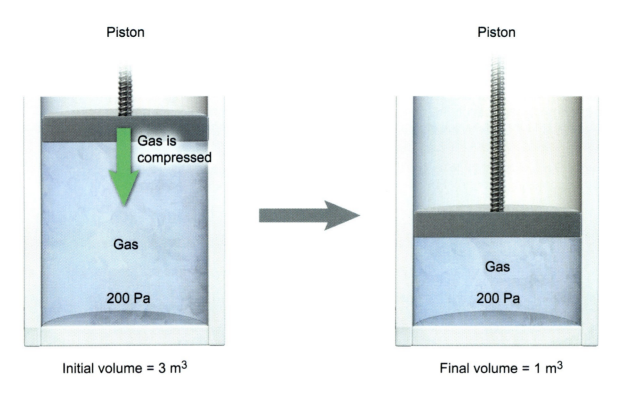

Figure 5.10 A piston compresses a gas at constant pressure.

Because the piston does work on the gas, it tends to increase the gas's internal energy. In the physics convention, W should be negative; indeed, in terms of the final and initial volumes:

$$W = P \cdot (V_{\text{final}} - V_{\text{initial}})$$

Inserting the values into this equation yields:

$$W = (200 \text{ Pa}) \cdot (1 \text{ m}^3 - 3 \text{ m}^3)$$
$$W = (200 \text{ Pa})(-2 \text{ m}^3)$$
$$W = -400 \text{ Pa} \cdot \text{m}^3 = -400 \text{ J}$$

In the first law, this negative work is associated with positive changes in the internal energy of the gas:

$$\Delta U = Q - W = Q - (-400 \text{ J}) = Q + 400 \text{ J}$$

By the chemistry convention, the work done is +400 J but the plus sign in the first law yields the same ΔU. With any convention, increasing the volume of the gas at constant pressure ($\Delta V > 0$) decreases its internal energy whereas decreasing volume ($\Delta V > 0$) increases its internal energy (Figure 5.11):

$$\Delta V > 0 \Rightarrow \Delta U < 0$$
$$\Delta V < 0 \Rightarrow \Delta U > 0$$

Work done by piston (surroundings) on the gas (system)

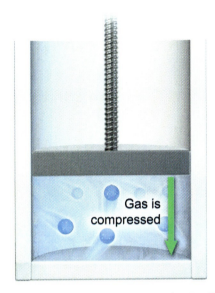

Volume V of gas decreases ($\Delta V < 0$)
Internal energy U of gas increases ($\Delta U > 0$)

Work done by gas (system) on the piston (surroundings)

Volume V of gas increases ($\Delta V > 0$)
Internal energy U of gas decreases ($\Delta U < 0$)

Figure 5.11 Work, volume, and internal energy for a gas in a container.

 Concept Check 5.4

A total of 500 J of heat flows out of a gas as it expands from 2 m³ to 5 m³ at a constant pressure of 500 Pa. What is the change in the gas's internal energy during this process?

Solution

Note: The appendix contains the answer.

5.2.03 Second Law and Entropy

When two systems are at different temperatures, a spontaneous transfer of heat must exist between them, as discussed in Concept 5.2.01. The **second law of thermodynamics** states that heat is spontaneously transferred from *a system of higher temperature to a system of lower temperature*. The reverse process, a spontaneous transfer of heat from lower temperature to higher temperature, never occurs in nature (Figure 5.12).

Chapter 5: Thermodynamics and Gases

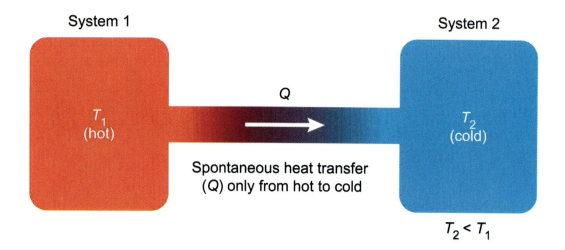

Figure 5.12 Second law of thermodynamics.

The second law reflects the observed fact that many processes in nature are **irreversible** (ie, proceed in only one direction). For example, when a sled moving on smooth ice hits rough ground, the friction force slows the sled to a stop, dissipating kinetic energy as heat. The reverse process, where heat flows from the ground-sled friction and speeds up the sled, does not occur.

Similarly, when a small container containing a gas is opened in a room, the gas particles spontaneously diffuse throughout the room (Figure 5.13). The reverse process, where gas particles in a room spontaneously move inside the small container, does not occur.

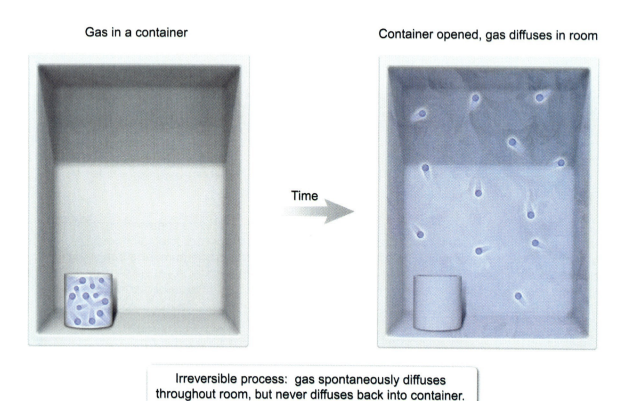

Figure 5.13 Diffusion is an irreversible process.

Irreversible processes are associated with the concept of **entropy** S, which is a thermodynamic **state function** (Concept 5.1.02) associated with the **disorder** in the system. Mathematically, entropy is equal to the product of the Boltzmann constant k_B and the logarithm of the number of **microstates** W:

$$S = k_B \log W$$

The Boltzmann constant is 1.38×10^{-23} J/K; hence, entropy has units of joules per kelvin. The number of microstates counts the number of ways the small-scale constituents of a system (eg, the molecules of gas) can be arranged to produce a given large-scale state (eg, the air in a room at some temperature), known as a **macrostate** (Figure 5.14).

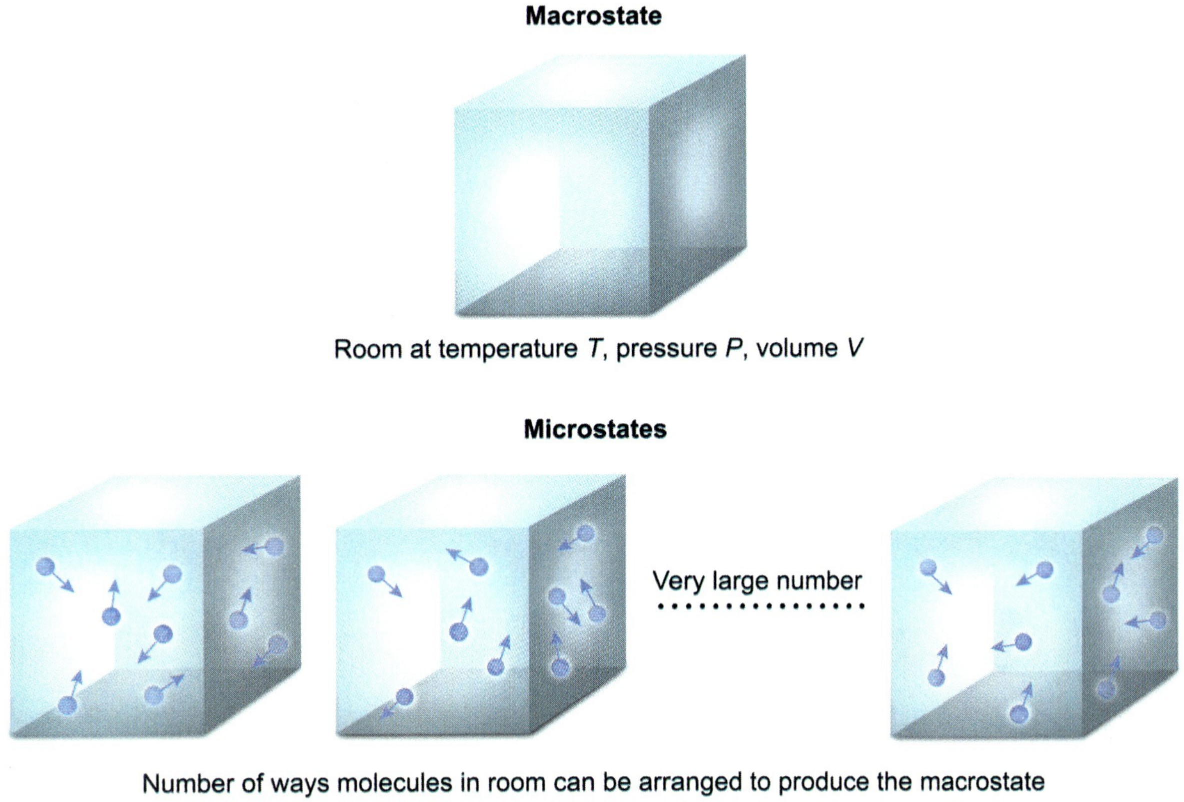

Macrostate

Room at temperature T, pressure P, volume V

Microstates

Very large number

Number of ways molecules in room can be arranged to produce the macrostate

Figure 5.14 A macrostate and its microstates.

The number of microstates measures the disorder of a macrostate. Roughly speaking, a system in a disordered state can be created in many ways, and therefore has many microstates and high entropy. Alternatively, not many ways are found to create an ordered system, so the number of microstates and the entropy are low (Figure 5.15).

Increased entropy yields increased disorder

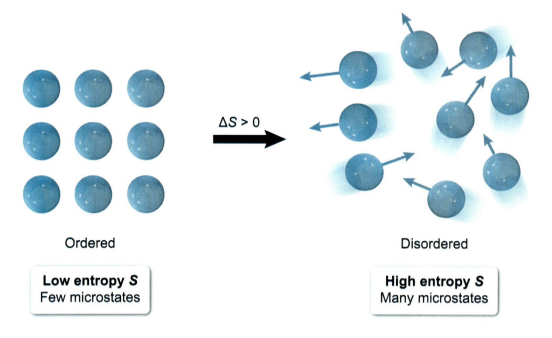

Figure 5.15 Entropy is associated with the disorder of a state.

In terms of entropy, the second law of thermodynamics states that the *entropy of an isolated system never decreases*; it must increase or remain constant in a process. Equivalently, the change in the entropy ΔS must always be greater than or equal to zero:

$$\Delta S \geq 0$$

Note that for closed and open systems, the entropy of the system can decrease. However, this decrease in entropy is compensated by a greater increase in the entropy of the surroundings:

$$\Delta S_{universe} = \Delta S_{system} + \Delta S_{surroundings} \geq 0$$

Hence, another statement of the second law is that the entropy of the *entire universe must always increase or remain constant*.

The second law is ultimately a statement about the **probability** associated with the macrostates of an isolated system. The probability that a system is in a disordered, high entropy state is much greater than the probability that the system is in a low entropy state. Therefore, it is extremely unlikely that a system spontaneously proceeds from a state of high entropy to a state of low entropy.

 Concept Check 5.5

Newton's laws of motion are reversible, which means the motion of the individual molecules in a gas is the same when time runs forward or backward. Why do the gas molecules that diffuse out of a container into a room not eventually move spontaneously back into the container?

Solution

Note: The appendix contains the answer.

Lesson 5.3

Heat

Introduction

This lesson focuses on heat, which represents the transfer of thermal energy to a substance or between two substances with different temperatures. Heat can be transferred from one object to another by conduction, convection, or radiation. Changing the temperature of an object causes its length to change depending on its coefficient of thermal expansion.

Furthermore, the calorimetry equation relates the amount of heat transferred to an object and its resulting change in temperature, which depends on the specific heat of the material or the heat capacity of the particular object.

5.3.01 Heat Transfer Mechanisms

Heat refers to the transfer of thermal energy into or out of any substance, whether solid, liquid, or gas. Specifically, two materials with different temperatures can exchange energy in the form of heat through three mechanisms: conduction, convection, or radiation. Heat energy is transferred from the warmer object to the cooler object until they reach **thermal equilibrium** and the temperatures of the two objects become identical (Concept 5.2.01).

Conduction

The **temperature** of an object depends on the average **kinetic energy** of its atoms. The object with a higher temperature has greater atomic kinetic energy. During heat transfer via **conduction**, some of the atomic kinetic energy of the warmer object is transferred to the cooler object as their atoms collide with one another. This transfer of molecular kinetic energy requires **direct physical contact** between the two objects, as shown in Figure 5.16.

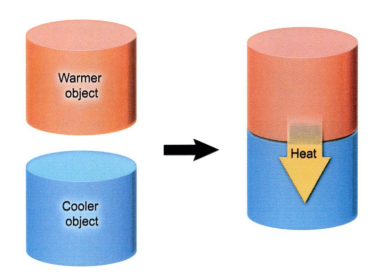

Heat transfer form warmer object to cooler object via conduction.
Objects are in physical contact

Figure 5.16 Heat transfer via conduction between two objects in direct physical contact.

Convection

During heat transfer via **convection**, a **flowing fluid** (ie, a liquid or gas) serves as a thermal intermediary and facilitates the transfer of heat. Some of the atomic kinetic energy of the warmer object is transferred to the flowing fluid, which is then physically transported to the cooler object. This transfer of molecular kinetic energy requires the two objects to be in contact with a common flowing fluid, but they are not in direct contact with each other.

Figure 5.17 demonstrates that convective heat transfer occurs when a furnace warms a room in a house. In this case, the flowing fluid is air, which is warmed in the furnace and transported into the cool room.

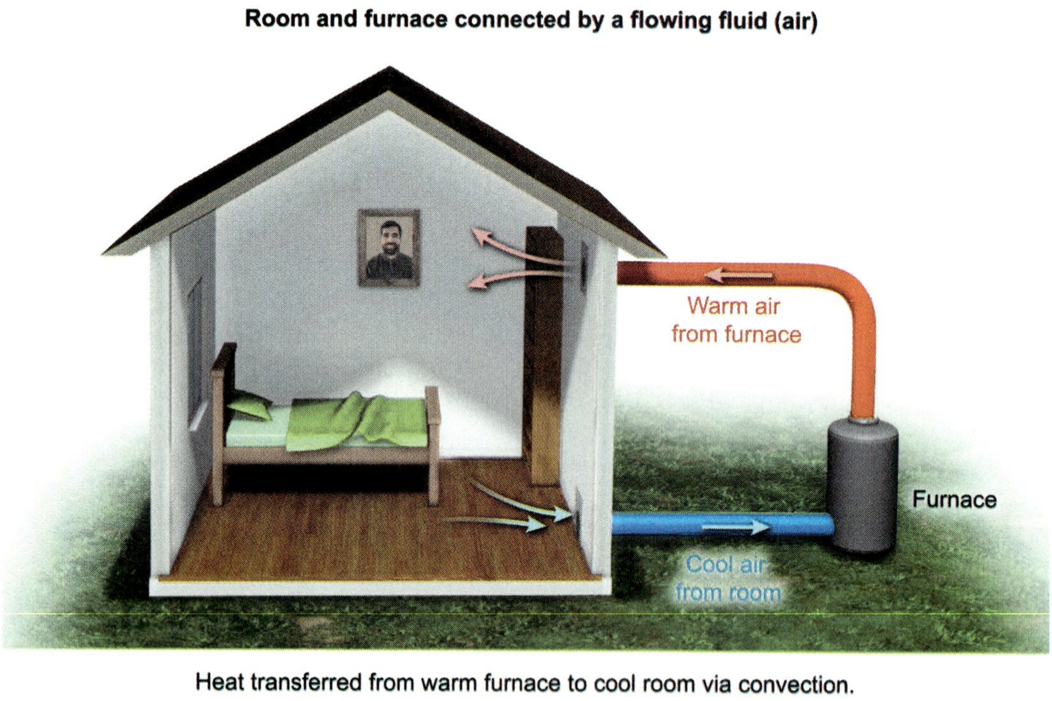

Room and furnace connected by a flowing fluid (air)

Heat transferred from warm furnace to cool room via convection.

Figure 5.17 Heat transfer from a warm furnace to a cool room via convection utilizes air as a flowing fluid.

Heat transfer through convection by means of a flowing fluid can be visualized when heating water on a stove. Figure 5.18 shows that the lower layer of water is warmed, becomes **less dense**, and rises to the top due to the buoyant force (Concept 2.1.03). This heated water is replaced by **cooler and denser** water from the upper layers, which is subsequently heated and then also rises. This upward movement of warmer water and downward movement of cooler water forms **convection currents** within the volume of water, leading to uniform heating and mixing of the water.

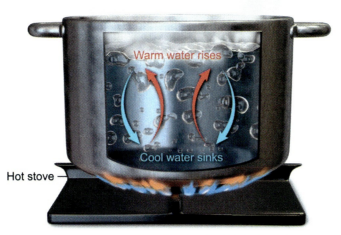

Heat transferred from bottom layer of water to top layer by convection

Figure 5.18 Heating water on a stove generates convection currents as warm fluid rises and cool fluid sinks.

Similar convective currents occur in Earth's atmosphere. The Sun heats Earth's surface and air near the surface, making the air less dense. As this **warmer surface air** rises, it is replaced with cooler and denser air from the **upper atmosphere**. These atmospheric convection currents help transport water vapor into the sky to form clouds.

Radiation

During heat transfer via **radiation**, the two objects are not in physical contact with each other or with a flowing fluid. Heat transfer by radiation relies on **electromagnetic waves**, which can carry energy across great distances. Some of the atomic kinetic energy of the warmer object is released as electromagnetic radiation, such as infrared photons, that travel to the cooler object and interact with its atoms, increasing their kinetic energy.

In fact, the two objects can be separated by large distances, or even by a vacuum. Heat transfer by radiation is the process by which Earth is heated by the Sun despite being separated by millions of miles and the cold vacuum of space.

Figure 5.19 demonstrates a common example of this heat transfer mechanism in the form of a microwave oven. Food placed in a microwave oven is exposed to electromagnetic radiation, which heats the food but does not directly heat the air inside the oven or the inner walls of the oven.

Heat transfer to frozen food via radiation
Frozen food exposed to microwave radiation and becomes hot food

Figure 5.19 Heat transferred to food via radiation using a microwave oven.

Physiological Heat Transfer

These heat transfer mechanisms are employed by living beings—including humans—to regulate body temperature. Blood flow in the skin and respiration are two important biological processes used by organisms to maintain thermoregulation.

Superficial blood vessels in the skin regulate blood flow by vasoconstriction and vasodilation. When the vessels dilate, blood flow increases and the temperature of the skin increases. The skin is in direct physical contact with the air, allowing greater heat transfer from the skin to the air via conduction (Figure 5.20). Similarly, constriction of the blood vessels reduces the skin temperature and decreases conductive heat transfer to the surrounding air.

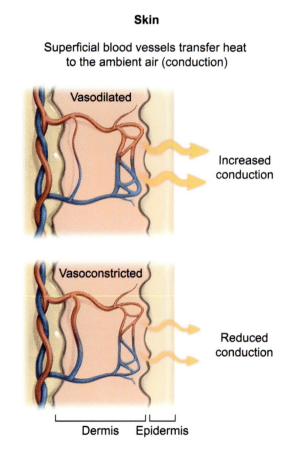

Figure 5.20 Transfer of body heat via conduction to the environment by blood flow in the skin.

Heat transfer to the environment through respiration occurs through similar mechanisms but relies on both conductive and convective heat transfer (Figure 5.21). Increasing the respiration rate or volume facilitates greater heat transfer into the environment whereas slower or shallower breathing decreases heat transfer.

Heat is efficiently transferred to the air inhaled into the lungs due to the large alveolar **surface area**. The air is in direct physical contact with the alveolar membrane and the blood vessels responsible for gas exchange. Hence, heat is transferred to the air in the lungs via conduction. This hot air within the lungs is then exhaled into the environment using the moving air as a flowing fluid to transfer the heat out of the body. Therefore, heat is transferred out of the body via convection.

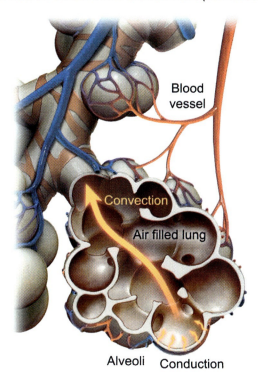

Figure 5.21 Transfer of body heat via conduction and convection through blood flow in the lungs and respiration.

 Concept Check 5.6

A customer orders a bowl of hot soup at a restaurant.

a. Which heat transfer mechanism is responsible for the increased temperature of the bowl when hot soup is poured inside?

b. Which heat transfer mechanism causes the soup to stay warm under an infrared heat lamp while it is waiting to be served to the customer's table?

c. Which heat transfer mechanism is involved when the customer blows on the soup to cool it down to a comfortable temperature?

Solution

Note: The appendix contains the answer.

5.3.02 Coefficient of Expansion

When an object is heated, it expands to a greater size. In cases of expansion in one dimension, the change in the object's length ΔL is equal to the product of its coefficient of thermal expansion α, the initial length L, and the change in temperature ΔT:

$$\Delta L = \alpha L \cdot \Delta T$$

The units of α are inverse degrees Celsius $\frac{1}{°C}$ (or inverse kelvins $\frac{1}{K}$). Different materials have unique values for α (see Table 5.1).

Table 5.1 Coefficient of thermal expansion for different materials.

Material	α(10⁻⁶/°C)
Aluminum	25
Copper	17
Glass	8.5
Iron	12
Silver	19

For example, a bridge spanning a river is 500 m long when the temperature is 10 °C, and it expands by 24 cm when the temperature is 50 °C (Figure 5.22). The thermal expansion equation can be rearranged to calculate the value of α for the bridge material:

$$\alpha = \frac{\Delta L}{L \cdot \Delta T}$$

$\Delta L = \alpha L \cdot \Delta T$

$\alpha = \dfrac{\Delta L}{L \cdot \Delta T}$

- **ΔL** Change in length
- **α** Coefficient of expansion
- **L** Length
- **ΔT** Change in temperature

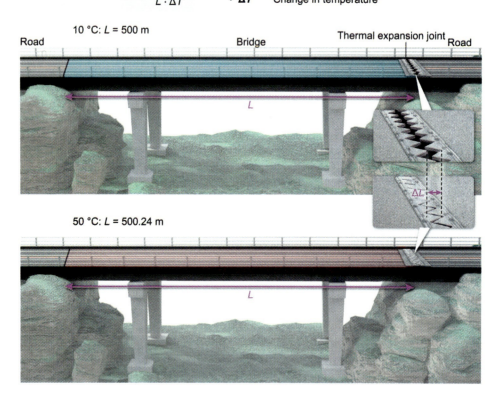

Figure 5.22 The thermal expansion of a bridge.

The ΔT equals the difference between 50 °C and 10 °C:

$$\Delta T = 50\ °C - 10\ °C = 40\ °C$$

Converting the units of ΔL from cm to m gives:

$$\Delta L = 24\ \cancel{cm} \left(\frac{1\ m}{100\ \cancel{cm}}\right) = 0.24\ m$$

Substituting these values into the equation for α yields:

$$\alpha = \frac{0.24\ \cancel{m}}{500\ \cancel{m} \cdot 40\ °C} = 12 \times 10^{-6}\ \frac{1}{°C}$$

> **Concept Check 5.7**
>
> Which material expands the least due to an increase in temperature: aluminum, glass, or silver? Assume the materials have the same initial length.
>
> **Solution**
>
> Note: The appendix contains the answer.

5.3.03 Heat Capacity and Calorimetry

The **calorimetry equation** is used to quantify heat exchange and temperature changes for different materials, including solids, liquids, and gases. The thermal properties of a particular material depend on its specific heat, which is an intensive property (ie, independent of mass) equal to the amount of heat required to increase the temperature of a unit mass (1 g) by 1 °C or 1 K. The calorimetry equation states that the total amount of heat Q gained or lost by an object is equal to the product of its mass m, the specific heat c, and the change in temperature ΔT:

$$Q = mc \cdot \Delta T$$

The only condition where this equation does not hold true is during a phase change. As discussed in Lesson 5.4, heat added to or removed from a substance during a phase change does not affect its temperature. On the exam, it can be assumed that no phase change occurs unless specifically stated otherwise.

The thermal properties of a specific object can also be described by its heat capacity, which is an extensive property (ie, dependent on mass). The heat capacity C is equal to the product of the object's mass m and the specific heat c for the material comprising the object:

$$C = mc$$

As a result, the calorimetry equation can be defined based on either the specific heat c of the material or the heat capacity C of a particular object:

$$Q = mc \cdot \Delta T = C \cdot \Delta T$$

The SI unit for specific heat is J/(g·°C), and the specific heat c_W of water is equal to:

$$c_W = 4.2 \, \frac{J}{g \cdot °C}$$

On the exam, the specific heat of a substance is generally provided. The one exception is c_W in units of **calories** (cal), which should be memorized as:

$$c_W = 1 \, \frac{cal}{g \cdot °C}$$

> **Concept Check 5.8**
>
> The heat capacity of a 10 g glass sphere is 8 J/°C. What is the specific heat of the glass?
>
> **Solution**
>
> *Note: The appendix contains the answer.*

The calorimetry equation can be rearranged to calculate the change in temperature from the mass, specific heat, and amount of heat Q gained or lost, yielding:

$$\Delta T = \frac{Q}{mc}$$

For example, sunlight delivers 8,400 J of heat to a birdbath containing 500 g of water with a specific heat of 4.2 J/(g·°C) and initial temperature of 15 °C. Substituting values into the equation above gives:

$$\Delta T = \frac{8,400 \, J}{(500 \, g)\left(4.2 \, \frac{J}{g \cdot °C}\right)} = \frac{8,400}{2,100} \, °C = 4 \, °C$$

Given that the temperature increases by 4 °C, the final temperature T_f is the sum of the initial temperature T_i and ΔT:

$$T_f = T_i + \Delta T = 15 \, °C + 4 \, °C = 19 \, °C$$

Therefore, the final temperature of the water in the birdbath equals 19 °C.

The calorimetry equation can also be applied when mixing two materials that have different temperatures. When two substances A and B are mixed, heat transfers from the warmer substance to the cooler substance until they reach thermal equilibrium. Due to conservation of energy, the heat Q_A gained by A must be equal but opposite to the heat Q_B lost by B, and vice versa:

$$Q_A = -Q_B$$

Replacing heat with the product of mass m, specific heat c, and change in temperature ΔT yields:

$$m_A c_A \cdot \Delta T_A = -m_B c_B \cdot \Delta T_B$$

For example, Figure 5.23 shows that 100 g of methanol with a specific heat of 2 J/(g·°C) and a temperature of −5 °C is combined with water with a heat capacity of 500 J/°C. The mixture reaches thermal equilibrium at a temperature of 20 °C. The heat Q_M gained by the methanol is calculated from its initial temperature T_i and final temperature T_f:

$$Q_M = m_M c_M \cdot \Delta T_M = m_M c_M \cdot (T_f - T_i)$$

$$Q_M = (100 \text{ g})\left(2 \frac{\text{J}}{\text{g} \cdot °\text{C}}\right)(20 °\text{C} - (-5 °\text{C})) = (100)(2)(25) \text{ J} = 5{,}000 \text{ J}$$

Conservation of energy implies that the heat lost by the water Q_W must be equal to $-Q_M$:

$$Q_W = -Q_M = -5{,}000 \text{ J}$$

The change in temperature of the water ΔT_W is calculated from its heat capacity C_W and Q_W:

$$Q_W = C_W \cdot \Delta T_W$$

$$\Delta T_W = \frac{Q_W}{C_W} = \frac{-5{,}000 \text{ J}}{500 \frac{\text{J}}{°\text{C}}} = -10 °\text{C}$$

Furthermore, ΔT_W depends on the initial temperature T_i of the water and final temperature T_f of the water, which is 20 °C at thermal equilibrium:

$$\Delta T_W = T_f - T_i$$
$$T_i = T_f - \Delta T_W = 20 °\text{C} - (-10 °\text{C}) = 30 °\text{C}$$

Therefore, the initial temperature of the water was 30 °C.

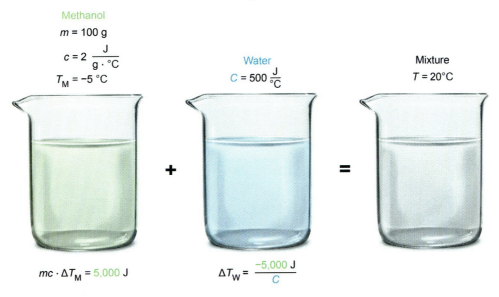

Figure 5.23 A mixture of methanol and water reaches thermal equilibrium.

Note that the mass of methanol was included in the calculation because its specific heat was provided. Conversely, the mass of water was not required because the heat capacity of the actual amount of water was given. Finally, note that the final temperature of the mixture is between the initial temperatures of the two substances but does not necessarily equal the average temperature of the two substances before mixing.

 Concept Check 5.9

Which requires more heat, increasing the temperature of 10 g of water from 20 °C to 40 °C, or increasing the temperature of 2 g of water from 20 °C to 80 °C? Assume the specific heat of water is equal to 4.2 J/(g·°C).

Solution

Note: The appendix contains the answer.

Lesson 5.4
Phases of Matter

Introduction

Ordinary matter exists as either a solid, a liquid, or a gas. This lesson introduces the characteristics of the three phases of matter, the processes by which matter transforms from one phase to another, and how to represent those transformations in a phase diagram as functions of pressure and temperature. The lesson concludes with a discussion of the energy flows related to the transformations between solid and liquid and between liquid and gas.

5.4.01 Phase Diagrams

The phases (or states) of matter describe the uniform physical state of the matter's molecules, and include solid, liquid, and gas (Figure 5.24):

- Molecules in a **solid** are tightly packed in an ordered structure and retain a fixed shape.
- Molecules in a **liquid** move past each other (but stay close together) and assume the shape of their container at a fixed volume.
- Molecules in a **gas** (or **vapor**) move past each other, can be well separated, and assume the shape and volume of their container.

Solid
- Particles closely packed
- Strong particle attractions
- Vibrational motion only
- Definite, fixed shape
- Specific volume

Liquid
- Particles in close contact
- Significant particle attractions
- Translational motion
- No definite fixed shape
- Volume similar to solids

Gas
- Particles far apart
- Minimal particle attractions
- Highly translational motion
- No definite fixed shape
- No definite volume

Figure 5.24 The three phases of matter.

As shown in Figure 5.25, **phase transitions** (ie, the transformation from one state to another) occur when energy is transferred into or out of a sample. The transition from solid to liquid is known as **melting**, and the reverse process (from liquid to solid) is known as **freezing**. Similarly, the transition from liquid to gas is known as **vaporization**, and the reverse process is known as **condensation**. Finally, the transition from solid to gas is called **sublimation**, and from gas to solid is called **deposition**.

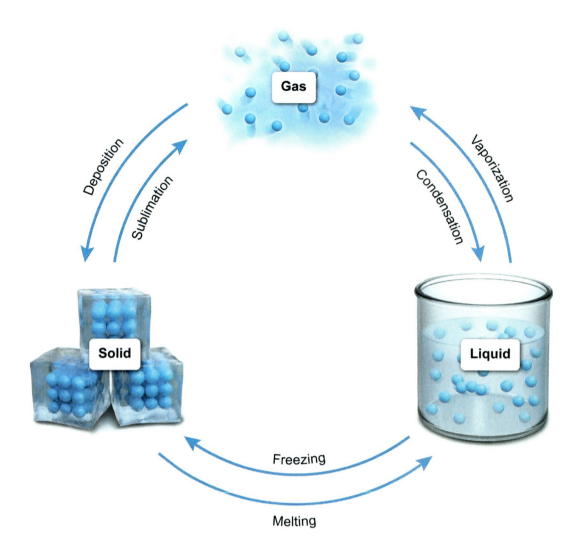

Figure 5.25 The phases of matter and the transitions between them.

The phase of matter depends on the temperature and pressure of its environment, and therefore changes in pressure and/or temperature can cause changes in phase.

Furthermore, **entropy** measures the **disorder** of a system (Concept 5.2.03). Solids have relatively low entropy because the structure of the molecules is ordered, and their motion is more constrained. Gases have relatively high entropy because the molecules are disordered, and their motion is less constrained.

Entropy changes as matter transitions from one phase to another. Changing from a solid to a liquid (melting) increases entropy but changing from a gas to a solid (deposition) decreases entropy (Figure 5.26).

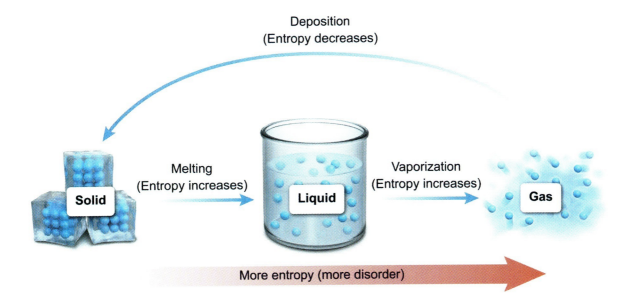

Figure 5.26 Entropy changes as a substance changes phase.

A **phase diagram**, like the one shown in Figure 5.27, illustrates a substance's stability in each phase (solid, liquid, or gas) as a function of **pressure** and **temperature**. The **boundary lines** between phases indicate the conditions under which two phases are in equilibrium. If a change in temperature or pressure results in the **crossing of a boundary line**, the substance undergoes the corresponding change in phase.

The **triple point** is the point at which all three boundary lines intersect, and it indicates the temperature and pressure at which all three phases simultaneously exist in thermodynamic equilibrium. Every pure substance has a unique phase diagram.

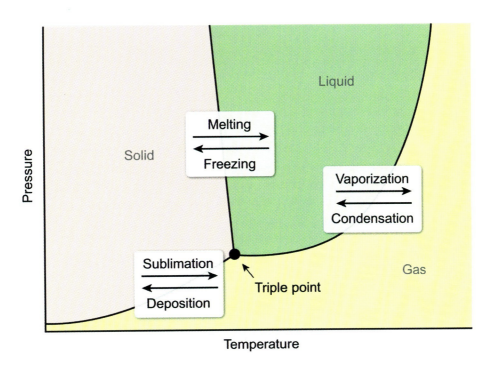

Figure 5.27 A phase diagram shows the phase transitions on a graph of temperature vs. pressure.

Water is a unique substance in that all three of its phases exist over a small range of pressures and temperatures. Figure 5.28 illustrates that a sample of water at a pressure of 1,000 Pa and a temperature of 265 K exists in the solid state (ice). Increasing the temperature by adding heat and maintaining a constant pressure of 1,000 Pa results in the solid transforming into liquid (ie, melting) at a temperature a little below 273 K. Once the sample is all liquid, continuing to add heat at the same constant pressure results in the liquid transforming into a gas (ie, vaporizing) at about 290 K.

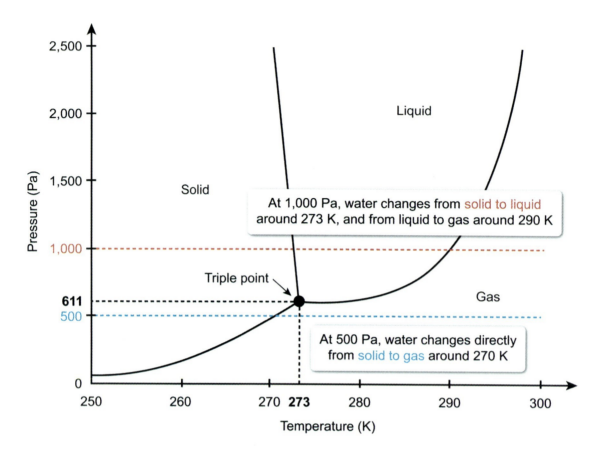

Figure 5.28 The phase diagram for water.

However, the same solid sample of water at a constant pressure of 500 Pa transforms directly from solid to gas (ie, sublimates) at a temperature of about 270 K. Further adding heat simply increases the temperature of the gas; no liquid state is possible for water at pressures below 611 Pa.

It can also be observed from the phase diagram that the **triple point** of water is a temperature of 273 K and a pressure of 611 Pa. Under these conditions, water exists simultaneously as ice, water, and gas. In other words, at that point water freezes, melts, and boils at the same time.

Concept Check 5.10

Atmospheric pressure is about 1 atm (101 kPa) at sea level, and the pressure decreases with increasing altitude above Earth's surface. According to the phase diagram shown for an unknown substance, how do the freezing point and boiling point temperatures for the substance change with an increase in altitude?

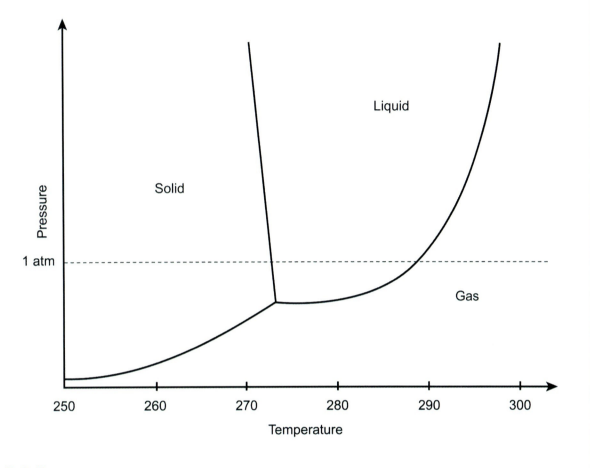

Solution

Note: The appendix contains the answer.

5.4.02 Heat of Fusion and Heat of Vaporization

As discussed previously, ordinary matter exists in three main phases: solid, liquid, and gas. When a substance changes from one phase to another (a **phase transition**), it absorbs (or releases) heat energy and its **temperature remains constant** until the entire substance has transitioned to the new phase, as shown in Figure 5.29.

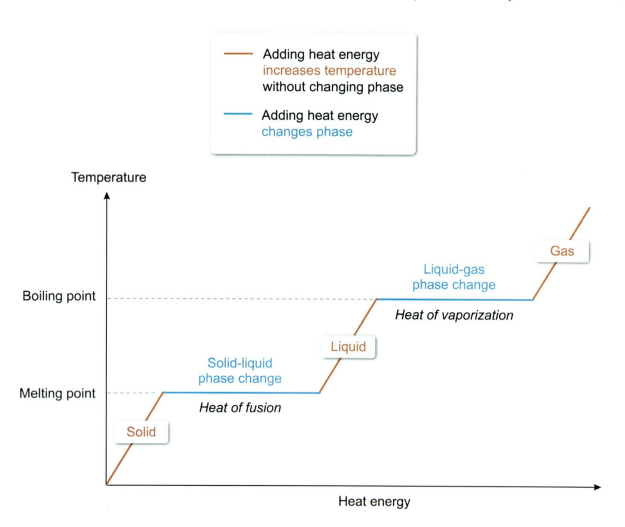

Figure 5.29 Phase change diagram.

For example, a solid substance (eg, ice) increases in temperature as heat is added, as shown by the initial part of the graph in Figure 5.30. The relationship between the amount of heat added and the resulting change in temperature is described in Concept 5.3.03.

However, when the temperature of the solid reaches the **melting point**, additional heat no longer causes a change in the temperature. Instead, the heat added to the system breaks the attractive intermolecular forces between the solid molecules, causing the solid to enter the liquid phase. This heat is called the **latent heat of fusion**.

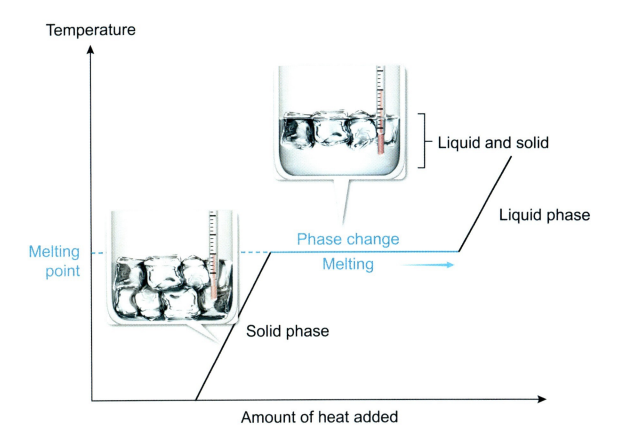

Figure 5.30 The latent heat of fusion transforms a solid into a liquid.

Removing heat energy from a liquid sample at the melting point temperature causes the *reverse* transformation from liquid to solid (ie, freezing) and liberates an amount of energy equal to the latent heat of fusion.

Furthermore, adding heat to a liquid (ie, water) raises its temperature, as shown in Figure 5.31. The relationship between the amount of heat added and the resulting change in temperature is described in Concept 5.3.03.

When the liquid reaches its **boiling point**, the temperature once again no longer continues to change with additional heat. The heat that goes into the system breaks the attractive intermolecular forces between the liquid molecules, causing the molecules to escape the surface of the liquid and enter the gas phase. The amount of heat required to evaporate the liquid is known as the **latent heat of vaporization**.

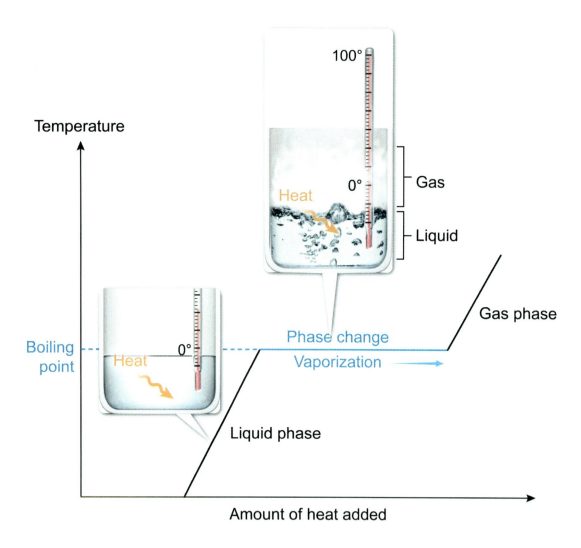

Figure 5.31 The latent heat of vaporization transforms a liquid into a gas.

Removing heat energy from a sample of gas at the boiling point causes the *reverse* transformation from gas to liquid (ie, condensation) and liberates an amount of energy equal to the latent heat of vaporization.

Concept Check 5.11

The phase change diagram for a sample of water initially at a temperature of −100 °C is shown, where the freezing and boiling point temperatures are 0 °C and 100 °C, respectively. In what phase(s) will the sample be after 10, 15, 40, and 70 kJ of heat are added?

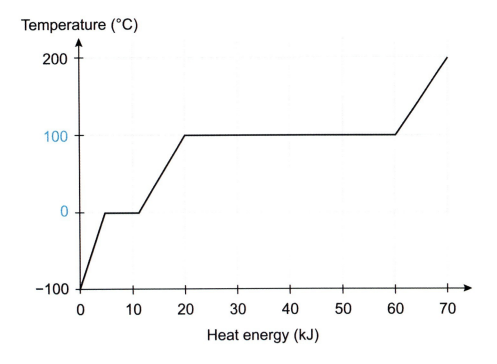

Solution

Note: The appendix contains the answer.

Lesson 5.5
Kinetic Theory of Gases

Introduction

This lesson begins by covering the concept of ideal and nonideal gases, which details the main differences between the scientific models used to approximate gas behavior and their special cases. Next, the kinetic theory for gases is introduced by offering a microscopic description of a gas and how this molecular view can be used to derive overall thermodynamic properties. Lastly, the lesson concludes with a discussion of the heat capacity of gases at constant volume or constant pressure.

5.5.01 Ideal and Nonideal Gases

Scientific models are often applied to describe complicated chemical or physical systems. By applying assumptions that simplify the system in question, a result can be obtained that is approximately the same as that observed in an experimental setting.

One such model is that of the **ideal gas**. An ideal gas is a *hypothetical* gas with characteristics that are consistent with certain simplifying assumptions made for systems involving real gases. An ideal gas has four characteristics (Figure 5.32):

- An ideal gas has **no attractive or repulsive forces** between the gas molecules.
- The size (ie, **molecular volume**) of individual gas molecules of an ideal gas is **negligible** (taken to be zero) compared to the volume of the container the gas occupies.
- **Collisions** between the molecules of an ideal gas are completely elastic (ie, no energy is lost by interactions or friction).
- Ideal gas molecules have an average kinetic energy (energy of motion) directly proportional to the **gas temperature**.

Chapter 5: Thermodynamics and Gases

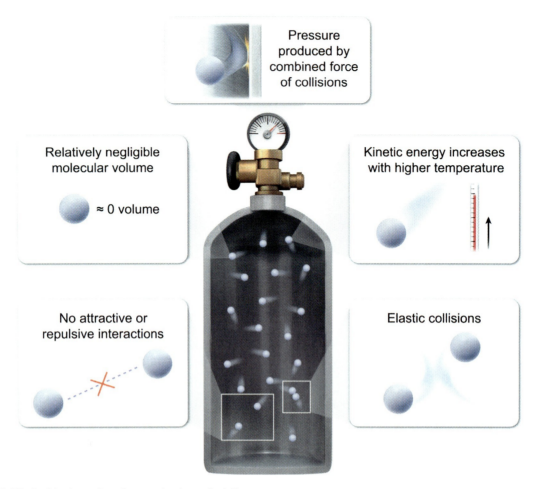

Figure 5.32 An ideal gas has four main characteristics.

Conceptually, these ideal characteristics assume that all the kinetic energy of ideal gas molecules manifests in container collisions to produce pressure, and that none of the kinetic energy is lost due to other interactions. Thus, systems involving real gases with minimal molecular interactions can be described reasonably well by this model.

The **ideal gas law** states that the product of the gas's pressure P and volume V is directly proportional to the product of the number of gas molecules N and the temperature T in kelvin:

$$PV = Nk_\text{B}T$$

where k_B is a proportionality factor called Boltzmann's constant (k_B = 1.38 × 10^{-23} J/K). The ideal gas law is known as an **equation of state** because it relates a system's **state functions** (Concept 5.1.02): pressure, volume, number of molecules, and temperature. For more details about the ideal gas law, including applications, see General Chemistry Concept 6.3.01.

The ideal gas model is a useful simplification of real gases, but it is not always an accurate representation of gases in the real world. Gases that do not conform to the ideal gas model are called nonideal gases and include situations where intermolecular forces are present, where the molecular volume is nonnegligible, or when collisions between molecules are nonelastic.

For example, at high pressures and low temperatures, the intermolecular forces in a gas become significant and the behavior becomes nonideal (Figure 5.33). In such situations the ideal gas law is no longer valid, and the gas can be modeled instead by the **Van der Waals equation**, which is covered in more detail in General Chemistry Concept 6.3.03.

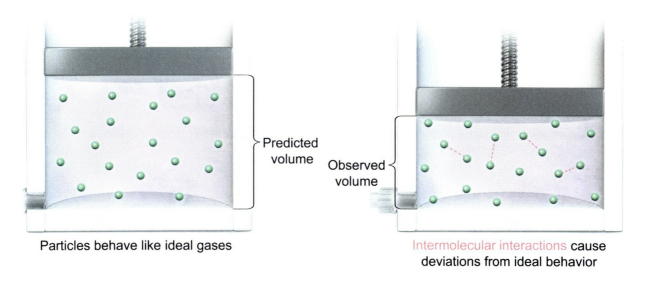

Figure 5.33 A gas under high pressure cannot be easily modeled as an ideal gas.

> ☑ **Concept Check 5.12**
>
> An ideal gas fills a container with a set pressure, a set volume, and a temperature that can be varied. Assuming the volume remains constant, how do the properties of the gas change if the temperature increases?
>
> **Solution**
>
> *Note: The appendix contains the answer.*

5.5.02 Kinetic Theory

The **kinetic theory of gases** is a model that explains the behavior of gases at the molecular level. According to the kinetic theory, gases are composed of many small particles (eg, atoms or molecules) in constant motion. These particles collide with each other and with the walls of the container, and the kinetic theory model is used to explain the macroscopic properties of gases (such as pressure, volume, and temperature) that arise due to these effects.

The key concepts that define the kinetic theory of gases are similar to those that define an ideal gas (Concept 5.5.01). However, two additional main concepts are necessary that relate to the pressure and volume:

- The pressure of a gas is related to the frequency and force of collisions between particles and the walls of the container, which are considered immovable. When the volume of the container decreases, the particles have less space to move around, causing them to collide with each other and the walls more frequently and resulting in an increase in pressure.
- The volume of a gas is related to the total space between all the particles. When the temperature is constant, an increase in pressure results in a decrease in volume (and vice versa).

Mathematically, the kinetic theory can be used to show that the product of pressure and volume is related to the number of molecules N, the molecule mass m, and the square of the average molecular speed $\bar{v}$:

$$PV = \frac{1}{3}Nm\bar{v}^2$$

Using the ideal gas law, it follows that:

$$PV = Nk_BT$$

$$\frac{1}{3}\cancel{N}m\bar{v}^2 = \cancel{N}k_BT$$

Hence, the average kinetic energy $\overline{KE}$ of ideal gas molecules is directly proportional to the product of k_B and the temperature T (in kelvin), as shown in Figure 5.34:

$$\overline{KE} = \frac{1}{2}m\bar{v}^2 = \frac{3}{2}k_BT$$

$$T \propto \frac{\Sigma \text{ KE of each molecule}}{\text{Quantity of molecules}} \qquad \begin{aligned} &\bullet T \quad \text{Absolute temperature} \\ &\bullet KE \quad \text{Kinetic energy} \end{aligned}$$

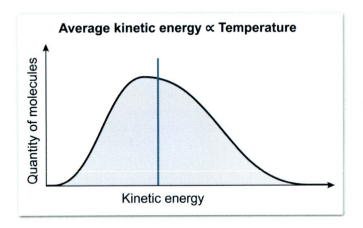

Container of gas molecules

Figure 5.34 Average kinetic energy as a measure of absolute temperature.

✅ Concept Check 5.13

An ideal gas within a container has an average molecular kinetic energy of 3.5 × 10⁻²¹ J. What is the approximate temperature of the gas in Celsius? (For Boltzmann's constant, use k_B = 1.4 × 10⁻²³ J/K.)

Solution

Note: The appendix contains the answer.

5.5.03 Heat Capacity at Constant Volume and Pressure

When heat is added to a substance, the average kinetic energy of the atoms or molecules within it increases. Consequently, **kinetic theory** implies that the temperature of the substance also increases. However, the thermal properties of the substance itself dictate exactly how the temperature changes per joule of heat energy exchanged.

Recall from Concept 5.3.03 that the **heat capacity** C is the ratio of the given heat exchanged Q and the temperature change ΔT:

$$C = \frac{Q}{\Delta T}$$

For ideal gases, the internal energy U depends on the temperature of the gas. Consequently, two types of heat capacity account for whether the thermodynamic process occurs at a constant volume or at a constant pressure. In particular, the heat capacity at a constant volume C_V is *not* equal to the heat capacity at a constant pressure C_P.

At **constant volume** V, zero work is done and the **first law of thermodynamics** (Concept 5.2.02) implies that the change in internal energy ΔU of the gas is equal to the heat energy:

$$\Delta U_V = Q = C_V \cdot \Delta T$$

However, at **constant pressure** P, the change in internal energy ΔU depends on both heat and the work done W because the ideal gas law states that a temperature change from a heat transfer must produce a change in volume ΔV under constant pressure:

$$\Delta U_P = (Q - W) = (Q - P \cdot \Delta V)$$
$$\Delta U_P = C_P \cdot \Delta T - P \cdot \Delta V$$

Note that solids and liquids also experience changes in volume when heat is added at constant pressure, but the work done by the molecules in solids and liquids is negligible. Hence, in these cases C_P and C_V are approximately the same.

If the internal energy change (ie, temperature change) in an ideal gas is assumed to be constant between constant volume and constant pressure conditions, the expression for ΔU_V can be substituted into the equation above to give:

$$\Delta U_V = \Delta U_P$$
$$C_V \cdot \Delta T = C_P \cdot \Delta T - P \cdot \Delta V$$

Solving for C_V gives:

$$C_V = C_P - \frac{P \cdot \Delta V}{\Delta T}$$

Using the ideal gas law yields:

$$C_V = C_P - \frac{Nk \cdot \cancel{\Delta T}}{\cancel{\Delta T}}$$
$$C_V = C_P - Nk$$

Hence, the heat capacity at constant volume is always *less than* the heat capacity at constant pressure (Figure 5.35):

$$C_V < C_P$$

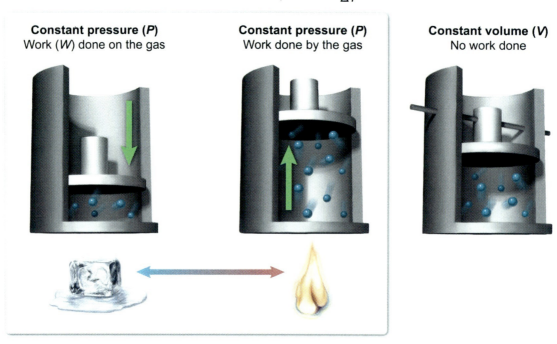

Figure 5.35 Heat capacity at constant pressure and volume.

Concept Check 5.14

A sample of gas with a known heat capacity at constant volume C_V and constant pressure C_P is heated from 25 °C to 50 °C in two different experiments. In the first experiment, the gas is placed in a closed, rigid container with a volume of 2.0 L. In the second experiment, the gas is placed in a 2.0 L container but the container can be compressed and stretched such that the pressure remains at a constant 1.0 atm. To calculate the amount of heat energy required to raise the temperature of the gas in each experiment, which heat capacity should be used in each case?

Solution

Note: The appendix contains the answer.

END-OF-UNIT MCAT PRACTICE

Congratulations on completing **Unit 5: Thermodynamics**.

Now you are ready to dive into MCAT-level practice tests. At UWorld, we believe students will be fully prepared to ace the MCAT when they practice with high-quality questions in a realistic testing environment.

The UWorld Qbank will test you on questions that are fully representative of the AAMC MCAT syllabus. In addition, our MCAT-like questions are accompanied by in-depth explanations with exceptional visual aids that will help you better retain difficult MCAT concepts.

TO START YOUR MCAT PRACTICE, PROCEED AS FOLLOWS:

1) Sign up to purchase the UWorld MCAT Qbank
   IMPORTANT: You already have access if you purchased a bundled subscription.
2) Log in to your UWorld MCAT account
3) Access the MCAT Qbank section
4) Select this unit in the Qbank
5) Create a custom practice test

Appendix

Concept Check Solutions

You will find detailed, illustrated, step-by-step explanations of each concept check in the digital version of this book.

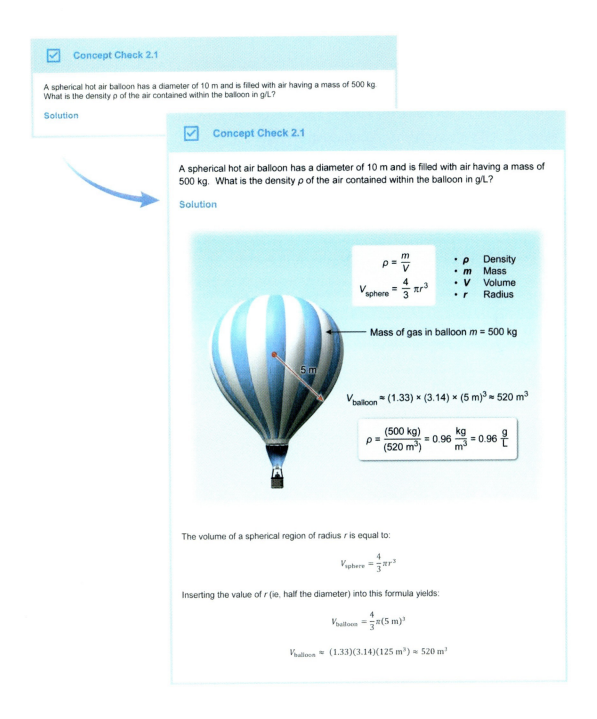

In this section of the print book, you will only find short answers to the concept checks in each chapter. Please go online for an interactive and enhanced learning experience.

Unit 1. Mechanics and Energy

Chapter 1. Motion, Force, and Energy

Lesson 1.1

1.1 6×10^{-9} m

1.2 Length^3 / (Time^2 × Mass)

1.3 (6 m, −8 m) and 10 m

Lesson 1.2

1.4 (a) 100 m, (b) 0 m, (c) 0 m/s, (d) 2 m/s

1.5 Velocity is directed downward. Acceleration is directed upward.

1.6 Time for the pebble to reach the bottom of the well.

1.7 $(2 \times v_i^2 \times \cos\theta_0 \times \sin\theta_0)/g$

Lesson 1.3

1.8 10,000 N

1.9 Trial 1: $a = 15$ m/s^2, Trial 2: $F_{net} = 50$ N, Trial 3: $m = 3$ kg

1.10 (a) Nail strikes the hammer, (b) Bag exerts a force on the boxer, (c) Moon pulls on Earth, (d) Rope pulls up on the student, (e) Sink pulls down on the student

1.11 $a_A < a_B$

1.12 All three free-body diagrams only have the gravitational force acting on the ball in the downward direction.

1.13 $F_N = F_g - F_T \times \sin\theta$, F_N increases as θ decreases

1.14 0.1

1.15 −2 m/s^2

1.16 16 N/m

Lesson 1.4

1.17 g, $4mg$

1.18 Top left

1.19 Each valve experiences the same torque.

1.20 x-direction: $F_g = F_f/(\sin\theta)$, x-direction: $F_g \times \cos\theta = F_N$ (Note, a rotated coordinate system is used with the x-axis parallel to the incline)

1.21 Shifts toward the clubhead

Lesson 1.5

1.22 (a) 0 J, (b) 0 J

1.23 2 m/s

1.24 (a) 0 J, (b) −24 J

1.25 ≈ 9 m/s

1.26 36,000 W

1.27 (a) ≈ 590, (b) 1,200 m

Unit 2. Fluids

Chapter 2. Fluid Dynamics

Lesson 2.1

2.1 0.96 g/L

2.2 5,000 g, 1.25 g/cm³, and 1.25

2.3 16,000 Pa

2.4 4 cm²

2.5 The total density of the boat is less than the density of water.

2.6 0.9 kg/L

2.7 The surface tension of the water decreases.

Lesson 2.2

2.8 16 times greater

2.9 5.4 kPa

2.10 4.9×10^{-5} m³/s

2.11 4×10^{-3} m/s

Unit 3. Electrostatics and Circuits

Chapter 3. Electricity and Magnetism

Lesson 3.1

3.1 +2Q

3.2 $F_{-2Q} = -2F_{+Q}$, F_{+Q} acts upward and to the right. F_{-2Q} acts downward and to the right.

3.3 3,000 N/C, Force acts to the right.

3.4 7.3×10^6 m/s

Lesson 3.2

3.5 4.0 mA

3.6 Reduce the wire's length by a factor of 4 or increase the wire's cross-sectional area by a factor of 4.

3.7 15 V

3.8 20 W

3.9 31 Ω

3.10 (a) 12 V, (b) 12 V, (c) 5 A, (d) Increases, (e) No change

3.11 Ammeter in series with R_2 and voltmeter in parallel with R_2.

Lesson 3.3

    3.12   50 mC

    3.13   9.9 J

    3.14   50 μF

    3.15   Parallel

Lesson 3.4

    3.16   Choice C and Choice D

    3.17   Toward the ground

    3.18 $(mv)/(eB)$

Unit 4. Light and Sound

Chapter 4. Waves, Sound, and Light

Lesson 4.1

    4.1 $A = 2$ m, $\lambda = 8$ m, $T = 4$ s, $f = 0.25$ Hz

    4.2   Up and down, and side-to-side

    4.3   0.4 m/s

    4.4   Constructive interference at points *A* and *D*. Destructive interference at points *B*, *C*, and *E*.

Lesson 4.2

    4.5   (a) Tornado siren, (b) Both sounds have the same speed.

    4.6   50 dB

    4.7   By increasing the tension on the string.

    4.8   8 nodes and 9 antinodes

    4.9   Toward

    4.10   1,250 m/s

Lesson 4.3

    4.11 7.5×10^{14} Hz

    4.12 4.97×10^{-19} J

    4.13   Choice B and Choice D

    4.14   22°

    4.15   Choice C

    4.16   The diffraction pattern narrows.

    4.17   10.000002 GHz

Lesson 4.4

    4.18   Convex mirror

    4.19   Behind the mirror, and smaller than the object.

    4.20   Same side as the object, and larger than the object.

    4.21   Diverging lens, −13.3 D, 0.6

    4.22   (a) 10 D, (b) −9/5

    4.23   25 cm

    4.24   Nearsightedness

Unit 5. Thermodynamics

Chapter 5. Thermodynamics and Gases

Lesson 5.1

    5.1   Closed system

    5.2   (a) Work, (b) Adiabatic or isobaric

Lesson 5.2

    5.3   −270.15 °C

    5.4   −2,000 J

    5.5   Entropy does not spontaneously decrease.

Lesson 5.3

    5.6   (a) Conduction, (b) Radiation, (c) Convection

    5.7   Glass

    5.8   0.8 J/(g·°C)

    5.9   Increasing the temperature of 10 g of water from 20 °C to 40 °C.

Lesson 5.4

    5.10   Freezing point increases and boiling point decreases.

    5.11   Solid and liquid phase, liquid phase, liquid and gas phase, gas phase

Lesson 5.5

    5.12   Pressure increases and the gas may display non-ideal behavior.

    5.13   −106 °C

    5.14 C_V for the first experiment and C_P for the second experiment.

Index

A
absolute zero, 245
absorption, 202–3
acceleration, 15–23, 29–32, 40–41, 63
adhesive forces, 95, 97
adiabatic process, 243
air resistance, 41, 67
ammeter, 137–38
amperes, 125, 152
amplitude, 161–63, 167, 179, 185, 198
angle of incidence, 204, 206–7
angle of reflection, 204
angle of refraction, 206–7
antinode, 184–86, 188, 191
Archimedes' principle, 90–91, 93
atmospheric pressure, 87–88, 269
attenuation, 177
average acceleration, 16
average force, 71–72
average power, 71–72
average speed, 12
average velocity, 10–11, 13–14, 21, 245

B
barometer, 87–89
battery, 128, 130, 132, 135–36, 139
Bernoulli's equation, 99–100, 102
blood flow, 104–9, 129, 196, 258–59
blood pressure, 106
blood velocity, 101, 109
blood vessels, 100, 102, 104–6, 258
Boltzmann's constant, 253, 276, 278
buoyant force, 90–91, 256

C
calories, 262
calorimetry, 261–62
capacitor, 139–45, 147
capillary action, 95, 97
Celsius, 245–46, 278
center of mass, 47, 50
charge, 115–23, 126, 131, 139–43, 149–50, 152–55
chromatic aberration, 230–31
circuit, 128, 130–32, 134–38, 144–45
circular polarization, 210
closed system, 102, 241

coefficient of friction, 43–44
coefficient of thermal expansion, 259–60
cohesive forces, 95
collisions, 35, 275–77
combination of lenses, 228
concave lens, 222–23
concave mirror, 218–20, 223
condensation, 265, 272
conduction, 255, 258–59
conductors, 126
conservation of charge, 116, 133
conservation of energy, 66, 123, 134, 203, 248, 262–63
constructive interference, 167–68, 185, 211, 213, 215
contact forces, 27
convection, 255–56, 258–59
converging lens, 224–32, 236
convex mirror, 218, 220–21, 223
cornea, 232
Coulomb's law, 117–18
critical angle, 206
current, 125, 134–35, 137–38

D
decibels, 180–80, 183–84
density, 83–84, 92, 94–95, 187, 189–90
deposition, 265–66
destructive interference, 168–69, 184, 211–13
dielectric constant, 143
diffraction, 211, 214
diopters, 226, 228
disorder, 253–54, 266
dispersion, 206–7, 230
displacement, 9–11, 14, 20–21, 44–45, 57–59, 61–62, 64, 69, 71, 162, 184
diverging lens, 224–26, 229–31, 234–35
Doppler effect, 191–95, 216
dynamic equilibrium, 54–55

E
elastic force, 44–45
elastic potential energy, 64
electric field, 19–23, 125, 127, 140, 142, 199, 209
electric potential energy, 123, 131, 203

electromagnetic radiation, 199–203, 207–9, 214, 257
electromagnetic spectrum, 199–201
electromagnetic waves, 164, 199, 201, 257
electromotive force, 132
electrostatic force, 117–21, 123, 140
entropy, 251, 253–54, 266–67
equilibrium, 44, 64, 68, 93, 162, 242, 245–46, 267

F
farsightedness, 235–36
first law of thermodynamics, 248, 279
floating object, 92–94
fluid, 83–95, 99–102, 104–5, 108, 127, 249
focal length, 218, 224–30, 234
focal point, 197, 217–19, 222–24
free-body diagram, 37, 39–41, 51, 53–54
freezing, 265
frequency, 161–63, 166–67, 174–78, 185–86, 188–89, 191–203, 216, 230, 277
friction, 37, 41, 43–44, 64, 104, 275

G
gases, 241–80
gravitational acceleration, 63, 66, 86–88, 90, 102–3
gravitational force, 22, 28, 37–38, 52, 60, 65, 123

H
harmonics, 185, 188
heat, 241–43, 248–49, 251–52, 255–58, 261–64, 268–71, 278–79
heat capacity, 261–64, 278–80
heat transfer mechanisms, 255–59
Hooke's law, 44, 161
hydrostatic pressure, 84, 86–87, 90

I
ideal fluid, 99–102, 104, 106, 108
ideal gas, 275–79
ideal gas law, 276, 278–79
images, 68, 195, 217–21, 224–30, 233–36
inclined plane, 40, 54, 75
incompressibility, 99
index of refraction, 206–7, 230
inertia, 27–30, 166
infrared, 200
instantaneous acceleration, 16–17
instantaneous speed, 11
instantaneous velocity, 11, 14, 16

insulators, 126, 143
intensity, 179–80, 182–83, 201, 203
intensive properties, 261
interference, 167–70, 195, 211
intermolecular forces, 41, 95, 276
inverted image, 218, 224
irreversible process, 252
isobaric, 243
isochoric, 243
isolated system, 241, 254
isothermal, 243

J
joules, 57, 60, 63, 69, 140, 253, 278

K
kelvin, 175, 245–46, 253, 276, 278
kinetic energy, 60–61, 63, 66–68, 102, 203–4, 245, 257, 276
kinetic friction, 42–43, 58, 65, 67
kinetic theory of gases, 277
Kirchhoff's laws, 125

L
laminar flow, 100, 104–5
latent heat of fusion, 270–71
latent heat of vaporization, 271–72
lens, 221–32, 234–35
lens aberration, 228
lens strength, 226, 236
lever, 72, 74–75, 77
lever arm, 50
linear kinematics equations, 18, 21, 23–25
liquid, 83–84, 95–97, 176–77, 255–56, 265–69, 271–72
logarithm, 180, 253
longitudinal wave, 164–65, 173
Lorentz force, 152–55

M
macrostate, 253–54
magnetic field, 149–50, 152–55, 199
magnification, 227–29
mechanical advantage, 51, 72–77
mechanical energy, 71
mechanical waves, 161, 166, 173, 179
medium, 161–62, 164–67, 173–77, 205–6, 221
melting, 265–66, 268
metals, 127–28, 173
microstates, 253
mirrors, 217–21, 223, 230

N

nearsightedness, 234–35
net force, 27–32, 40–41, 49, 51, 54
net work, 248–49
Newton's first law, 27
Newton's second law, 30
Newton's third law, 33
node, 184, 188, 191
nonconservative forces, 64, 67
nonideal fluids, 99–100, 104–5
normal force, 36–38

O

ohmmeter, 137
Ohm's law, 128–32, 135, 138
open systems, 241, 254
optical instruments, 221, 228, 230
oscillations, 161–62, 164, 173, 207

P

parallel circuit, 132–33, 135
parallel force, 75
parallel light rays, 218, 222
parallel plate capacitor, 139, 142–43
pascals, 84, 104
Pascal's law, 88
period, 161–63, 166–67, 174–75
perpendicular force, 84
phase diagram, 267–69
phase transitions, 265, 267, 269
phase (waves), 168–69, 209–10, 212
phases of matter, 265–69, 273
photoelectric effect, 203–4
photon, 201–2
pitch, 174
Planck's constant, 201, 203
Poiseuille's law, 104–5, 108, 130
polarized light, 207–10
potential energy, 63–64, 66, 102, 125
power, 69–71, 131–32, 179, 182, 201
pressure, 84–90, 99, 102–3, 105, 107–8, 242–43, 249, 265–69, 276–78, 280
probability, 254
process functions, 243–44
projectile motion, 22–25
pully, 72
PV loop, 107–8
Pythagorean theorem, 6

R

radiation, 199–200, 208, 257
radius of curvature, 218
real images, 218–19, 221, 223–24, 226–27, 230
reflection, 177, 195, 202, 204, 206, 217
refraction, 204–7, 221, 230
resistance, 104, 127–30, 135, 137
resistivity, 126–28
resistor, 128–39, 144
resonance, 184–85, 195
resonant frequency, 185, 187, 198
retina, 232–36
Reynolds number, 108–9

S

scalars, 3, 5
second law of thermodynamics, 251–52, 254
series circuit, 133–34
shock waves, 197–98
sign conventions, 249
simple machines, 72, 75, 77
slope, 13–14, 16–17, 45
Snell's law, 206, 221
solids, 176, 261, 265–66, 279
sound intensity, 179, 183
sound waves, 165, 174–79, 182, 195–97
specific gravity, 84
specific heat, 261–64
speed of light, 199, 205
speed of sound in air, 189, 198
spherical aberration, 230–32
spherical mirror, 218, 220–21, 223–24, 226
spring, 44, 64, 68–69, 161
spring constant, 45, 68
standing waves, 184–86, 188–89, 191, 195, 198
state functions, 241–42, 253, 276
static equilibrium, 41, 51–52, 54, 93
static friction, 41–43, 59
sublimation, 265
submerged object, 90–94
surface area, 43, 95–97, 182, 258
surface tension, 95–97
surroundings, 95, 241, 248–49, 254
system, 47–48, 51, 63, 72–73, 116, 228–30, 241–43, 245–49, 251, 253–54, 275–76
system of lenses, 229

T

temperature, 175, 241–48, 251, 253, 255, 258, 260–71, 276–80
tension, 37–39, 166, 187, 195

thermodynamic equilibrium, 242, 246, 267
thermodynamic systems, 241, 248
thermometer, 245
thin lens equation, 226–28, 234
thin lenses, 221, 223–24, 227
torque, 50–51, 54–55
total internal reflection, 206
transverse waves, 164, 207
triple point, 267–68
turbulence, 104

U
ultraviolet, 200
unit analysis, 122
upright image, 217, 224, 233

V
van der Waals equation, 276
vaporization, 265, 269, 271–72
vasoconstriction and vasodilation, 258
vector components, 6
vector quantities, 3, 5, 10, 27, 117, 119, 150
velocity, 5, 9–11, 13–18, 22, 27, 30, 60–61, 66, 69, 104–5, 152, 154
Venturi effect, 102–3
virtual image, 217–20, 224–27, 230
viscosity, 99, 104–5, 109, 129–30
visible light, 200–201
voltage, 121, 123, 125–26, 128–43
voltmeter, 137–38
volts, 123, 140
volumetric flow rate, 101

W
wavelength, 161–67, 174–77, 185–86, 188, 190–91, 199–203, 207, 211–14, 216
weight, 22, 29, 37, 40, 51–52, 86, 93, 96
work, 57–65, 69, 77, 102, 108, 122–23, 179, 248–51, 279
work-energy theorem, 60–62
work function, 203

X
X-rays, 201–2, 214

Z
zeroth law of thermodynamics, 245–47